ISBN 978-1-332-30542-1
PIBN 10311759

1 MONTH OF
FREE
READING

at

www.ForgottenBooks.com

By purchasing this book you are eligible for one month membership to ForgottenBooks.com, giving you unlimited access to our entire collection of over 1,000,000 titles via our web site and mobile apps.

To claim your free month visit:

www.forgottenbooks.com/free311759

English
Français
Deutsche
Italiano
Español
Português

www.forgottenbooks.com

Mythology Photography **Fiction**
Fishing Christianity **Art** Cooking
Essays Buddhism Freemasonry
Medicine **Biology** Music **Ancient
Egypt** Evolution Carpentry Physics
Dance Geology **Mathematics** Fitness
Shakespeare **Folklore** Yoga Marketing
Confidence Immortality Biographies
Poetry **Psychology** Witchcraft
Electronics Chemistry History **Law**
Accounting **Philosophy** Anthropology
Alchemy Drama Quantum Mechanics
Atheism Sexual Health **Ancient History**
Entrepreneurship Languages Sport
Paleontology Needlework Islam
Metaphysics Investment Archaeology
Parenting Statistics Criminology
Motivational

OUTLINES

OF

ANATOMY AND PHYSIOLOGY,

TRANSLATED FROM THE FRENCH

OF

H. MILNE-EDWARDS,

DOCTOR OF MEDICINE, PROFESSOR OF NATURAL HISTORY AT THE ROYAL
COLLEGE OF HENRY IV., AND AT THE CENTRAL SCHOOL
OF ARTS AND MANUFACTURES, IN PARIS.

———

By J. F. W. LANE, M. D.

———

BOSTON:
CHARLES C. LITTLE AND JAMES BROWN.
———
MDCCCXLI.

BOSTON:
PRINTED BY FREEMAN AND BOLLES,

TRANSLATOR'S PREFACE.

THE subject of Anatomy and Physiology, until quite recently, was never presented to the scholar, or to the public in general. Previous to that period its study had been limited to medical men only, and a certain degree of opprobrium even was attached to those engaged in Anatomical pursuits. But such a change has been gradually wrought in public opinion upon this subject, that now no system of education is considered complete, without some acquaintance with the human form and structure be included. The prejudice is fast passing away, nay has already passed, which has so long kept back from the many a knowledge of the healthy exercise of their organs and functions; and which no doubt in a great degree prevented more rapid advances from being made by physicians themselves in the intimate structure of the organs. Upon the importance of this study to the medical student it were needless to enlarge — the information obtained from it finds daily use with all, whether in diet, ventilation, clothing, broken bones, running, walking: not an act of the body can be performed without calling in question some point of physiology. But let us bear in mind, that the present is strictly a physiological view of man, and no where does our author so much as present a glimmer even of his psychological state. This limited view may however serve as a stepping stone to the other; and from this point how vast, how illimitable the field of Natural History appears.

There are many excellent works upon the same subject, but they are too minute and abound in technicalities; to supply the deficiency arising from this cause, and at the same time furnish the medical student with an elementary work, has been my object in this translation. In the language of the author, " the age and characters of some of my readers has, perhaps, made it necessary for me to pass rapidly over certain functions; but I am not aware of any important omissions in this sketch of the phenomena constituting life. My aim has been to be clear and concise, and from the nature of my work I have been compelled to limit myself to facts, and to pass over in silence the opinions and hypotheses, with the discussion of which our treatises on physiology are so filled." It was my intention to have added a series of notes, but upon a reperusal of the work they do not seem so necessary as at first; the few remarks which have been appended will not, I trust, be found wholly useless.

TABLE OF CONTENTS.

B

ANATOMY AND PHYSIOLOGY.

PRELIMINARY REMARKS.

THE object of these lessons is to make known _{these lessons.} ^{Object of} the most interesting points in the history of animals, to signalize those which are useful or injurious to man, and to show their respective influence upon industry and riches in general.

To do this, the principal phenomena, which characterize their mode of existence, must first be exposed, and their form and structure described ; the means indicated, by which naturalists distinguish between them, and recognise with certainty all these beings, the number of which is so great as nearly to terrify the imagination. It must be shown, how they live, how they are distributed upon the different parts of the surface of the globe, how they contribute to the well-being of man, or prove noxious to his industry, of what riches they are the source, and of the means to which we have recourse to procure them, or to draw from them and appropriate to our own desires a portion of the products.

they furnish. In other words, these lessons will be devoted to the teaching of comparative anatomy and physiology.

The simple announcement of the subject on which I propose to treat, will suffice to make its interest and importance appreciated. In truth, where is the man, whose curiosity has not been a thousand times excited by the sight of the singular animals, which people all around, and which present at every instant of their lives, phenomena so remarkable, and, often, so incomprehensible? Who, when reflecting upon these phenomena, and the actions we ourselves execute, does not experience the desire to scrutinize the interior of these complicated machines, to know by what agents, by what mechanism all these movements are executed; to examine the processes, by which all organized beings assimilate to their own substance the foreign substances by which they are nourished; and to inquire the uses of the different organs, of which the body is composed. In short, every enlightened man must perceive, that the history of animals producing pearls, silk, and wool, which furnish us our daily food, or lend us their power, is of no mean importance; and no one can remain indifferent to the knowledge of a multitude of other animals, which, if less useful, are not less interesting, from the wonderful instinct with which nature has endowed them.

It would then be useless to stop here, to prove that the study of zoölogy is a necessary element of education, or to establish by examples the interest presented to us in this branch of human knowledge; and I shall therefore hasten to considerations more worthy to engage our attention.

GENERAL CHARACTERS OF LIVING BEINGS.

When we cast our eyes over the immense _{Anatomical and physiolog-} number of animals, which populate the surface _{ical characters common to an-} of the globe, we are at first only struck by the _{imals and veg- etables.} enormous and innumerable differences they present ; a man, a fish, a spider, and an oyster, for example, to a superficial observer have nothing in common. But if we examine with care these different beings, we shall soon be convinced, that notwithstanding these differences, there are a certain number of characteristics possessed by each of them, and reproduced without exception in all the other animals. These characteristics are of two kinds ; some are furnished us by the material disposition of the body, and consequently belong to *anatomy* ;[1] the others consist in the phenomena presented by these same bodies during life, and they belong to *physiology*,[2] a science, which treats of the actions and properties of living bodies.

The chief distinction between animals and vegetables, and all the other bodies of nature, is life, an interior movement, the cause of which is unknown, but whose effects are easily perceived.

All living beings, and these alone, possess _{Nutrition.} the faculty of duration for a time, and under a particular

---

[1] Anatomy is that branch of the natural sciences, which treats of the form, position, structure and qualities of the organs composing the bodies of living beings. This name is derived from the Greek word ἀνατομία, whose roots (ἀνὰ within and τέμνειν to cut) indicate the manner, in which anatomical studies must be made.

[2] From its etymology (φύσις nature and λόγος discourse) the word physiology should signify discourse upon nature or science of nature, but it is only employed in the signification given above.

form, by drawing constantly to their composition, appropriating to themselves, a part of the surrounding substances, and rendering to the exterior world a part of their own; in other words, these beings possess the faculty of self-nourishment; and, when the constant addition of materials, of which they are composed, stops without return, then this body dies, and is soon completely destroyed; now, this movement has always a limited duration, and death is a necessary consequence of life.

With brute bodies, such as stones and minerals, it is entirely different. Once formed, they exist, so long as a foreign force does not destroy them; and during this time, not necessarily limited, they are not the seat of a movement of nutrition: if their volume augments, it is by the simple juxtaposition of another body similar to themselves; and if they lose a part of their own substance, it is by the action of a force from without, and entirely independent of the cause of their existence.

The continual movement of composition and decomposition, constituting the nutritive function of which living bodies are the seat, escapes our senses; but its existence is revealed to us by numerous facts, very readily perceived.

Proofs of the existence of the nutritive movement. The desire constantly experienced by animals to introduce into the interior of their bodies foreign substances, which may serve them as food, already suffices for the presumption, that these beings must continually incorporate with their own organs materials drawn from without; and only by these means can the increase of volume, so remarkable in all these beings during the first periods of their existence, be satisfactorily explained. A new-born infant weighs but about six pounds, and twenty-five years after, when it has become adult, its weight exceeds an hundred; plainly then, at this epoch of its life, it has already derived from substances, originally foreign, the greater part of the materials of which its organs are composed. On the other hand, the extreme emaciation, which ensues upon certain diseases, proves that the living body may abandon a portion of the matter, of which it was formed, and render to the external world a part of its own substance.

The experiments of Sanctorius, who, in order to study the phenomena of transpiration, passed a great part of his life in a balance, prove also that the human body constantly experiences diminutions of weight, which the aliments are destined to repair.

But a single example can scarcely fail to remove all doubt of the existence of the nutritive movement, even in the hardest and deepest parts of the body. An English surgeon, Belchier, having by chance eaten of a hog raised by a dyer, observed that the bones of the animal were red, and attributing this peculiarity to the fact, that it had been nourished on coloring matter, he conceived the possibility of using similar means to render visible the effects of the nutritive function; and the experiments undertaken by him at that time resulted in complete success, and have since been satisfactorily repeated by a great number of distinguished men. When animals have been nourished upon madder for a certain time, it is always found that the bones assume a red color; but if, after an animal has been thus fed, the use of the madder is suspended, in a definite time the red matter, which should have been deposited in the substance of these organs, is no longer found, and must therefore have been rejected. Now these facts can only be explained by the continual movement of composition, or decomposition, to which the name of nutrition has been given.

Duration and origin. We have already seen, that after a certain duration the nutritive movement is always stopped, and that all living beings, after having existed for a time, the extreme limit of which is fixed for each of them, must necessarily perish. But this destruction of the individual does not include the disappearance of the species; for another character, common to all these beings, is the faculty of producing other beings like themselves, or in other words, of perpetuating their race by the phenomena of generation.

The origin of organic differs completely from that of brute bodies. The latter have existed from the creation of the world, or are formed by the combination of other bodies in nothing resembling themselves. Living bodies, on the contrary, always arise from a being like themselves, from a parent, to which they are at first united, and from which they are separated, only when their development is so far advanced, that they can maintain an independent existence.

Structure. All living beings have a common structure. Their body is always formed by an union of parts dissimilar in themselves, some of them being liquid, others solid. It is a spongy tissue composed of lamina, or of solid and very extensible fibres, which leave between them interstices filled with fluid ; and this kind of structure, which has received the name of ORGANIZATION, is one of the essential conditions of their existence.

To give these bodies form, solid parts are evidently necessary; and to sustain the nutritive movement, to make penetrate into their intimate tissue the foreign bodies destined to be incorporated there, to throw off the particles ceasing to belong to them, fluid parts are also requisite. The latter should be present in every part in which life is to be sustained, should penetrate into the substance of the solids as well as be diffused upon their surface, and consequently these solid parts must have a spongy and areolar texture. It is therefore impossible to conceive of the existence of a movement similar to the nutritive, without a mode of structure like that we have described; and observation teaches us that this organization is met with in all living beings, animal, as well as vegetable. Thus to these has been given the name of *organic*, in opposition to brute beings, which are styled *inorganic bodies.*

Finally, the chemical nature of the materials, Chemical composition. of which they are made up, is equally characteristic. We always find in these bodies a certain number of substances also to be met with in the inorganic kingdom, and which here offer nothing peculiar; water for instance; but the products, which form the essential base of all the solid parts of living bodies, belong properly to the organic kingdom, and present very remarkable properties. The number of these substances is very considerable, and although differing much in themselves they are, for the most part, formed of the same elements united in different proportions; in general, compounds of carbon, hydrogen and oxygen, or of substances resulting from the union of these three elements with a fourth principle, called azote.[1]

Review of the characters of organized beings.   We see then, that there exist in nature two classes of bodies ; the organized or living bodies, and the brute or unorganized ; and, by reviewing what has been said upon their properties, we shall find that the general characteristics of the former, in which alone we are interested, consist in their *mode of structure and chemical composition, in the power they possess of nutrition and reproduction, in their origin, and in the limited duration of their existence.*

GENERAL CHARACTERS OF ANIMALS.

Chemical composition.   If we now restrict yet closer the field of our observations, and confine ourselves to the study of

---

[1] Those substances are called elements, or simple bodies by chemists, from which only particles of the same nature can be extracted : for example, iron, gold, and sulphur; at the present day about fifty are recognised, and by their various combinations all the other bodies in nature are formed. *Carbon*, pure and crystalized, constitutes the diamond; while uncrystalized and mingled with certain salts, it constitutes the ordinary charcoal; thus, to be convinced of the existence of this elementary principle in all organized bodies, it must be borne in mind, that when warmed by contact with air so far as to decompose them, the residue is charcoal. *Hydrogen*, so called because it enters into the composition of water, presents itself, when isolated, under the form of a gas or aeriform fluid of extreme lightness; combined with carbon, it forms the gas employed to give light, and is obtained by distillation from coal, oil, etc. *Oxygen* is also a gas ; it forms about the fifth part of the atmospheric mass, and produces by its combination with a great number of bodies the phenomena of combustion ; united with hydrogen it forms water ; and united to carbon, the carbonic acid gas, which is met with in the sparkling wines, beer, etc., and is also disengaged from burning charcoal. It is called oxygen (which means generator of acids) because it enters into the composition of the greater part of the acids. Finally, *azote* is a gas, which, combined with oxygen, constitutes the atmosphere.

animals, we shall find that these beings have also other characteristics, which are common to them, but not met with in the organized bodies of the vegetable kingdom. And first; the chemical composition of these beings is not the same; the substances which constitute plants, include little, or no azote, but have carbon as a basis; while in animals azote plays the principal part.[1]

In these latter beings too, life is manifested ^{Faculties more varied than in vegeta-bles.} in a much more complicated form than in veg- etables. To the faculty of nutrition and of reproduction, is added that of executing, under the influence of an inward volition, movements which tend to a determined end, and that of feeling, or receiving impressions, and being conscious of them. Whence has arisen the name of *animated beings,* given to animals in opposition to vegetables, which are called *inanimate.*

To define, then, concisely, the word *animal* ^{Definition of the word animal.} we may say, that it is applied to every body endowed with the faculty of nutrition, of feeling, and of executing spontaneous movements.

THE FUNCTIONS OF ANIMALS AND THEIR ORGANS.

The different phenomena by which life is ^{Organs.} manifested, are always the result of the action of some part of the living body; and these parts, which may be regarded as so many instruments, bear the name of *organs.* Thus an animal can execute movements only

---

[1] This difference in the composition of vegetable and animal substances is easily proved; when the latter are burned, the azote they include, combines with a certain quantity of hydrogen, and forms ammonia, which diffuses a peculiar odor when disengaged. Substances not containing azote do not possess this odor.

2

by the action of certain *organs* called muscles, and can
obtain a knowledge of what surrounds it only through
the intervention of the *organs* of the senses.

Apparatus.   When several organs concur to produce a
phenomenon, this assemblage of instruments is known
by the name of *apparatus*, and the action of an isolated
Function.   organ, or an apparatus is called the *function*.
For example, we say *locomotive apparatus*, to designate
the collection of organs which serve to transport the
animal from one place to another, and *function of locomotion*, to designate the action of all these parts.

Object of
different func-
tions.   two objects: the preservation of the individual
and the preservation of his race; but among the first
an important distinction remains to be established; some
serve for the support and increase of the body, others to
place the animal in relation with the beings which surround him.   As for the functions of reproduction, there
results from them the formation of new beings, similar
to those from which they originate.

Classification
of the functions.   The functions or acts of these beings may
then be divided into three great classes, viz. ; the *functions of nutrition, the functions of relation*, and the
functions of generation.   Those of nutrition and generation, as we have already seen, are common to plants
and animals; therefore to them has been given the col-
lective name of *functions of the vegetative life ;* but those
of relation exist only in the latter, and constitute what
physiologists call *animal life*.

Difference
between the
functions of
different ani-
mals.   some these acts are but few, and life is mani-

fested only by a small number of faculties; in others, on the contrary, the most varied phenomena may be observed, and the existence of a multitude of faculties, with which the former are not endowed. Now, each of these phenomena, as we have already seen, is the result of the action of some part of the body; and observation, as well as reason, prove that the mode of action of an organ, or instrument, depends always upon its intimate nature, structure, and its other properties. It follows, therefore, that the organization of different animals must offer as little uniformity as the modes, according to which these beings fulfil the three orders of functions already mentioned.

In animals with the most limited faculties and simplest life, the body every where presents the same structure. The parts composing it are all similar; and, the identity of organization presupposing an analogous mode of action, the interior of these beings may be compared to a work-shop, where the workmen are employed in the execution of similar labors, and where consequently their number would influence the quantity, but not the nature of the products. Each part of the body performs the same function as its neighbor, and the general life of the individual is composed only of phenomena, characterizing the life of some one of its parts.

But in proportion as we ascend in the series _{the vital labor.} of beings, as we approach man, the organization becomes more complicated; the body of each animal is composed of parts, becoming more and more dissimilar, as well in form and structure, as in function; and the life of the individual results from the accumulation of a constantly increasing number of instruments, possessed of different faculties. It is at first the same organ, which feels, moves, absorbs from without nutritive substances, and which continues the species; but by degrees the different functions are localized, they acquire instruments proper to themselves, and the different acts, of which they are composed, are executed in their turn by distinct organs. Thus, the more the life of an animal is compounded of different phenomena, and the more subtile its faculties, the greater is the division of labor in the interior of its body, and the more complicated its bodily structure.

The principle which seems to have guided nature in the perfection of beings, is, as we see, precisely the one which has had the greatest influence upon the progress of human industry : *the division of labor.*

Consequences of this law. There is another consequence of this law, which merits our attention, and which necessarily presents itself to the mind, however little we may meditate upon the facts of which we have just spoken. We have seen, that the higher an animal stood in the series of beings, the more various were the instruments destined to produce the collection of the vital phenomena, and the more special and limited were the functions of each of these organs. It results then, that the destruction of

Effects of mutilations. any part of the body must produce trouble in the economy, in proportion to the perfection of the faculties of the animal; and that those beings will remain so much the less injured by mutilations, whose bodily structure is less complicated.

Now this observation conduces to the explanation of several facts, which, at first, appear incomprehensible, and finds its confirmation in certain phenomena so extraordinary, that they would have been rejected as fables, if they had not been stated by men, whose testimony is indisputable.

Thus there exist animals, whose bodies may be divided into many pieces without arresting Experiments upon the poly- pi of fresh wa- ter. the vital movement; on the contrary, each fragment afterwards takes on an unusual development, and soon constitutes a new animal, similar in form to that from which it springs, quite as perfect in its kind, exercising the same functions, and living in the same manner.

Fig. 1.[1]

The singular beings, called by naturalists fresh water polypi, or hydræ, and which are often found under the water lentils, present this queer phenomenon; mutilate them as you will, far from killing you multiply them. A naturalist of Geneva, of the last century, Tremblay, to whom the knowledge of these curious facts is due, opened one of these little animals; then he stretched and cut it in various ways, completely chopping it up, and, notwithstanding this state of extreme division, each of its fragments soon became a complete animal.

[1] In the figure 1 are represented several hydræ attached to the water lentils (a); these animals, as will be shown, consist only of a small gelatinous tube, open at one of its extremities, and surrounded by a circle of filaments called *tentacula*, by the aid of which they introduce aliments into their digestive cavities. One of the polypi (b) bears on the sides of its body two young, which spring from it, and will soon be detached.

To comprehend this phenomenon, in appearance so contradictory to all that the higher order of animals teaches us, we must first examine the mode of organization of the polypi, of which we have but now spoken. These animals are too small to be profitably studied with the naked eye ; but when seen through a microscope, the substance of their body is found to be every where identical ; it is a gelatinous mass, in which no distinct organ can be perceived, enclosing globules extremely small.

Now, as we have already remarked, identity in the mode of organization necessarily supposes identity in the mode of action, of the faculties. It therefore follows, that all the parts of the body of our polypi, having the same structure, must fulfil the same functions ; each of them must unite in the same manner with the rest to the production of the phenomena, whose union constitutes life ; and the loss of one, or several, of these parts does not therefore presuppose the cessation of any of these acts. But if this be true, if each portion of the body of these animals is capable of feeling, motion, nutrition, and the reproduction of a new being, we see no reason, why, after being separated from the rest, it may not, if placed in favorable circumstances, continue to act as before ; and why each of these fragments of the animal may not only continue the performance of those functions, necessary for the support of life, but also reproduce a new individual, and perpetuate its race ; phenomena fully proved by the testimony of Tremblay.

Experiments upon earth worms. Let us now apply this same principle to beings, whose structure is less uniform, and whose different acts have already instruments appropriated to each of them. Let us take, for example, the lumbricus terrestris, or earth worm.

In this cylindrical and slender animal the localization of the functions has already been established; nutrition is composed of a series of acts executed by different instruments; digestion is effected in a cavity, whose walls possess particular properties; there exists also a system of canals, serving to conduct the nutritive materials to all parts of the body, and an apparatus, which has become the principal seat of the faculty of perceiving impressions, and determining movements; and finally, we find instruments intended merely for locomotion. Thus we cannot conceive the possibility of dividing in every way the bodies of these worms, as was done in the polypi, without death. But, if we examine the disposition of these different sets of apparatus, which concur, each in a different manner, to the support of life, we shall find that they extend uniformly from one extremity of the body to the other, and that each transverse segment of the animal differs but little, or not at all, from the others; it is a constant repetition, and represents to a certain extent the entire animal, for it includes all the organs, the action of which is necessary to the vital movement. We can then easily conceive the possibility of detaching a certain number of these segments from the rest of the body, without causing either section to lose any of the vital properties enjoyed by the entire individual; and this really takes place. If an earth worm be cut transversely into two, three, ten, or twenty pieces, each of its fragments may continue to live as the whole, and to constitute a new individual.

Localization
of the different
functions in
the superior
animals.
But by ascending yet higher in the scale of animated beings, the division of physiological labor is seen to augment ; the different functions become the appendages of so many particular sets of apparatus ; each of the acts resulting is executed by a special instrument, and these different sets of apparatus, instead of being uniformly distributed through the whole extent of the body, are lodged in different parts ; so that the loss of each portion of the body deprives the animal of some faculty, and produces in the economy a disturbance great in proportion to the importance of the faculty in the support of life.

Examples to
be chosen for
the study of
the functions.
In studying the different functions of animals, I shall be obliged to indicate the manner in which each of them is complicated and perfected by means of this division of labor ; but I shall only dwell upon the most important facts, and shall confine myself to the examination of the phenomena of life in beings, which in this relation occupy the summit in the series of animals.   In truth, it is, when each of these phenomena results from the action of a particular instrument, that the different acts of which the function is composed, may the more easily be observed, and the effects of life be better analyzed ; consequently the most complicated animals are the best known to anatomists and physiologists, and offer to us the greatest interest.

ORGANIC TISSUES.

Materials
forming the
organs.
The bodies of these beings include a considerable number of different organs ; but if the comparative structure of these different parts be examined, we shall soon be convinced that the materials, of

which they are composed, are far less variable than was at first supposed. They are in truth the same tissues, differently combined and taking upon themselves particular forms, which constitute the greater part of our organs. The principal organic tissues are three, namely, the muscular, nervous, and cellular.

The *muscular tissue* constitutes what is com- ^{Muscular tissue.} monly called the *flesh* of animals ; it is the productive agent of all their movements, and always consists of fibres capable of contraction. Sometimes these fibres, so to speak, are dispersed in the substance of our organs ; at others, they are collected into masses and form *muscles* ; but whatever be their arrangement, they are always distinguished by their contractile property ; and in the body of man, as well as of most animals, they are met with in those situations, where there are movements to be executed.

The *nervous tissue* is a soft, usually whitish ^{Nervous tissue.} matter, which constitutes the brain and nerves, and is the seat of the faculty of perception ; when treating of the functions of relation we shall have occasion to study its properties and uses.

Finally, the *cellular tissue*, so called from its ^{Cellular tissue.} spongy and areolar texture, is of all the materials constituting our organs the most universally diffused. In the simplest animals it appears to form almost the totality of the body ; and in those which have, like man, the most complicated structure, this tissue forms a layer more or less thick between all the organs, it fills the interstices these parts have between them, and is also met with in their very substance, where it serves to unite the different portions of which they are composed, as on the

3

surface it tends to connect the different apparatus of the economy ; it is in a measure the link of all the organs, and by its different modifications gives birth to the membranes, and to many other tissues ; finally, in it the fat is always deposited, but the little sacs containing this matter are completely distinct.

This tissue is a whitish, semi-transparent, and very elastic substance, composed of filaments and small lamellæ more or less consistent and irregularly connected, so as to leave between them grooves, or cellules of variable size. These cellules have but incomplete walls, and are separated from each other only by a kind of spongy lining; thus they communicate together, and afford a ready passage to the fluids, which traverse them ; lastly, they are moistened by an aqueous liquor loaded with albuminous particles, and known by the name of *serosity.*

The communication of these cellules is easily demonstrated ; if a hole be made in the skin of an animal recently killed, and air blown into the cellular tissue, this fluid penetrates to all parts of the body, and distends them. Butchers do this every day to make their meat look well ; it has also been practised by some beggars, to deform in a most hideous manner the bodies of unfortunate children, and thus excite curiosity, or public commiseration.

An example to the point. — A celebrated surgeon of the sixteenth century, Fabricius Hildensis, relates, that in 1593 there was exhibited at Paris an infant from fifteen to eighteen months of age, whose head was of monstrous size ; the parents of the little unfortunate carried it from city to city as an object of curiosity, and attracted a great number of spectators ; but a magistrate, having suspected some fraud, caused them to be arrested and put to the torture ; they then acknowledged, that a hole had been made in the skin upon the summit of the head of the child, and air blown in by means of a tube. The operation was repeated every day, and thus in time the head of their child had acquired this unusual aspect. In our day we have seen this barbarous practice pursued by a beggar of Brest.

The other organic tissues, which concur with the preceding to form the different parts of the body, are the serous and mucous membranes, the different varieties of the fibrous tissues, (tendons, aponeuroses, etc.) cartilages, bones, etc. ; but to all appearance these are only modifications of the cellular tissue. Finally, we often see them accidentally developed from the cellular tissue ; and in most of these cases the cause of their formation is known : thus, whenever the cellular tissue is submitted to pressure and constant friction, it is transformed to a serous membrane ; when in contact for some time with an irritating liquid, it puts on all the characters of the mucous membranes ; under the influence of traction and mechanical irritation, it gives birth to fibrous membranes : and it is to be remarked, that all these membranes exist in a normal manner in the economy, only where those causes act, which elsewhere would have determined their formation. The more accurate study of these tissues will naturally find its place in the course of this work ; we will only add here, that all, both the primitive cellular, the muscular, and the nervous tissues, appear in the final analysis to be composed of small globules, visible only by the aid of the microscope, and formed in circles of varying disposition.

Mucous, fibrous, bony tissues, etc.

Elementary structure of the tissues.

THE FUNCTIONS OF NUTRITION.

Nothing certain is known of the manner in which nutrition is effected; and it is even probable, that for some time yet the mechanism of this internal movement, the existence of which has been already demonstrated, will remain a mystery to physiologists; but if it has been impossible directly to observe the labor, by which the constituent materials of the organs are constantly renewed, greater success has attended the investigation of the different acts, which prepare, or accompany this curious phenomenon. We know the principal agent of nutrition, and how it is distributed to the different parts of the body. The manner, in which this agent, the blood, can transport to all the organs materials not originally belonging to it, but which were deposited in a certain part of the body, or simply in contact with certain parts, has been successfully studied. It has also been found, that in traversing the organs the blood is deprived of a portion of its constituent parts, gives birth to new liquids, and changes its nature to such a degree, that it is no longer suited to discharge its functions, until it has been, in some sort, regenerated by the action of the air; lastly, it has been seen, that the nutritious liquid, thus conducing to the support of the organs, is exhausted, and must be renewed from foreign matters suitably prepared, in organs especially appointed for this purpose.

These various phenomena of vegetative, or *organic life*, give rise to the functions of circulation, absorption, exhalation, secretion, respiration, and digestion; acts, which we are now about to study.

THE NUTRITIVE FLUIDS, OR THE BLOOD.

It has been shown that the nutritive function can take place only by the intervention of fluid parts, and that every organized body includes liquids as well as solids.

These liquids are merely water, holding in *Liquids contained in the body.* solution or suspension various substances, of which we shall speak hereafter. To their existence, in the thickness even of the solid parts of the body, animals owe in a great measure their rounded form, and from them their organs derive suppleness, and the other qualities necessary for the exercise of their functions. Thus by desiccation a tendon is diminished in volume, loses its suppleness, whiteness, and pearly brilliancy, and becomes hard, rigid, semi-transparent, and brownish ; but if then plunged into water, it will rapidly absorb this liquid, and retake in proportion to the absorption the properties it had lost.

From this it might easily be concluded, that *Effects of desiccation.* the desiccation of an organized body, carried to a certain degree, must always interrupt the vital movement, and produce death. And this is [always observed ; *Experiments upon the vibrio, etc.* but to demonstrate in a manner yet more evident the part these liquids play in the animal economy, I must state here the curious results obtained by Spallanzani, Buffon, Bauer, and some other naturalists in their experiments upon the desiccation of certain microscopic animalcules. When the water, in which their bodies are soaked, is in a great measure evaporated, many of these extremely minute beings lose the power of motion, and cease to afford any sign of life ; but they do not perish at once ; they may be preserved in this state of apparent death for a very long time, and to recall them completely to life only a little water is requisite ; this takes place in the *vibrio of wheat,* an animalcule resembling a small eel, or rather the ends of thread, and which lives in the grains of blighted wheat.[1]

If placed in a drop of water, and examined by the microscope, they are at first seen swimming with vivacity ; but when the liquid has evaporated, they remain immovable, and distil from their bodies a kind of varnish, which covers them and prevents farther desiccation. They are then completely deformed, and resemble in nothing living beings; notwithstanding, by plunging them into water they soon retake their forms, and return perfectly to life, even after having been in this state of apparent death for several months.

[1] Wheat is subject to several maladies, such as carbon, ergot, blight, rachitis, etc. The majority of these alterations depend upon the development of a species of insect, called *uredo,* in the substance of the grain ; but rachitis is owing to the presence of the vibrio, for this disease may be produced in ordinary wheat, by inoculating the grain with these animalcules. The vibrio is at the most from two to three lines in length, and about the sixth of a line in diameter.

Fig. 2.

Fig. 3.

Fig. 4.

Fig. 2 exhibits one of these animalcules as seen through a microscope; and Fig. 3 a grain of wheat magnified and cut in two, to display the vibrios lodged in it, and the alterations they have caused; lastly, Fig. 4 represents an ear of wheat, which has rachitis ; the black points are the diseased grains. What is said by Buffon with regard to the animalcules of ergoted wheat, must be applied to the vibrios of rachitised wheat, for in the ergot they are not met with.

Similar phenomena have been observed upon other microscopic animalcules, called by naturalists *Rotifera*, and met with in spouts. But with most animals it is quite different; for to them a real death is always the immediate consequence of desiccation to a certain degree. Fishes afford us a striking example, for when drawn from the water they soon perish, and their death is principally to be attributed to the desiccation of their gills.

The proportion of fluids contained in the body of an animal is much more considerable than we should have at first imagined. The tendons, of which we were speaking above, contain half their weight of water, and the proportion is much greater in the other organs. The body of a man contains about nine tenths of its weight of fluid; when a body, weighing an hundred and twenty pounds, had been dried in an oven seventeen days, its weight was found to be reduced to twelve pounds; and mummies have been found not weighing more than seven or eight pounds. In other respects the proportion of fluids and solids varies according to the animals, and individuals; we may generally say, that the relative proportion of the former is greater than the latter in youth, and animals of a simple structure. *Relative proportions of the fluids and solids.*

In the animals of an uniform structure, all the liquids of the economy are similar. They appear to be merely water, more or less loaded with organic particles; but in the beings occupying a more elevated rank in the zoölogical series, the humors cease to be all of the same nature, and there is one destined in a special manner to supply the wants of nutrition: this liquid is the *blood*. *Nature of the fluids. Blood.*

White blood. In the majority of inferior animals the blood is far from possessing those physical characters, by which it is recognised in man, and the animals nearly allied to him; instead of being red, and thick, it consists merely in an aqueous liquid, sometimes completely colorless, sometimes slightly tinged with yellow, rose, or lilac. It is therefore difficult to be seen; and for a long time it was supposed, that these beings were completely destitute of it, hence they were called *bloodless animals.*

The *animals with white blood,* or having the blood but slightly tinged, are very numerous : all the insects enter into this class, and it is an error to consider flies, as having red blood in their heads. When one of these animals is crushed, we see, it is true, a reddish liquid exuding, but this matter is not from the blood, it arises merely from the eyes of these little beings. Spiders, crabs, lobsters, and all animals resembling the latter, and which are classed by zoölogists as *crustacea,* have colorless blood; also snails, muscles, oysters, intestinal worms, and all other animals of the class *mollusca,* and of the *zoöphytes.*

Red blood. The blood, on the contrary, is red in all animals whose structure approaches man ; such as the mammiferæ, birds, reptiles, fishes, and even worms of the class, annelida.

Globules of the blood. When the blood of any of these beings is examined by the microscope, it is constantly found to be composed of two distinct parts ; of a yellowish and transparent liquid, to which has been given the name of *serum,* and of a multitude of little solid corpuscules, regular in form, and of a beautiful red color, which swim in the fluid just mentioned, and which are called the *globules of the blood.*

In man, and all other animals of the class Form of the globules.
mammiferæ (the dog, horse, ox, for example,) *Fig.* 5.[1]
the globules of the blood are circular; but in
birds, reptiles, and fishes, they have constantly *Fig.* 6.
an elliptical form. These corpuscules are extremely small. In man, the dog, hare, and
some other mammiferæ, their diameter is equal *Fig.* 7.
only to about the one hundred and fiftieth part
of a millimètre; in the sheep, horse and ox
they are but one two hundredth of a millimètre; in the
goat they seldom exceed the one three hundredth of a
millimètre. In birds the globules of blood are greater
than in the mammiferæ: their smaller diameter is in
general one one hundred and fiftieth of a millimètre, and
their greater diameter varies according to the animals,
from the one one hundredth to the one seventy fifth of a
millimètre. In the class of fishes and reptiles, these
corpuscules are yet greater; in the frog, for example,
their smaller diameter is one forty fifth of a millimètre,
and their larger one seventy fifth.

When these globules are attentively examined Structure of the globules.
by a powerful microscope, they are seen to be composed,
each of two distinct parts, consisting of a kind of bladder, or membranous sac, in the centre of which is found
a spheroidal corpuscule.

Usually this investment is depressed, and *Fig.* 8.[2]
forms around the central nucleus, a circular
rim varying in size, so that the whole pre- *Fig.* 9
sents the aspect of a lens. The exterior
envelope of the globules is formed of a kind

---

[1] Fig. 5, blood of a man; Fig. 6, blood of a sheep; Fig. 7, blood of a
sparrow. These globules are magnified one thousand times.

[2] Fig. 8, profile view of the globule in the blood of the frog, magnified

4

of jelly, easily divided, and of a variable red color; to
the presence of these vesicles the blood owes its color.
The central nucleus of the globules exhibits more consistence, and is not colored.

Coagulation of the blood. In the ordinary state the blood is always
liquid, and is composed, as we have already said, of an
aqueous fluid, holding solid globules in suspension ; but
there are circumstances, in which its physical properties
are completely changed. This takes place, for example,
every time that the blood is drawn from the vessels, in
which it is contained in the interior of the body of the
living animal ; when left to itself it is transformed in a
few moments to a mass of gelatinous consistence, which
is separated by degrees into two parts ; òne liquid, yellowish, and transparent, formed by the serum ; the other
more or less solid, completely opaque, and of a red color,
to which has been given the name of *clot*, or *fibrine of
the blood.* The latter is composed principally of globules
more or less changed.

The blood sometimes loses this property of coagulation.
This singular phenomenon is remarked in animals killed
by a strong electric shock, by lightning, for example, and
by the action of certain poisons, such as the bite of ven-
omous serpents. Lastly, at other times the blood may
resemble its usual mass, but afterwards it separates into
three parts, the serum, the clot, and a soft grayish layer,
which occupies the surface, and is called the *buffy coat.*
It is especially in persons affected with inflammatory
diseases, such as pneumonia, or inflammation of the
lungs, and acute rheumatism, that the blood is thus

about seven hundred times. Fig. 9, front view of the same ; the envelope
is torn so as to exhibit the central nucleus.

coated ; and the majority of physicians agree in regarding it as a certain sign of the existence of an internal
inflammation ; but recent observations prove, that the
formation of the coat may depend upon circumstances
entirely different, and which in themselves have no importance, such as the size of the opening in the vein, the
form of the vessel in which the blood is received, etc.

Chemistry teaches us, that the blood contains Chemical composition.
the greater part of the substances, which enter into the
composition of the different organs of the body which it
is destined to nourish. In man, to the hundred parts of
blood there are met with seventy-eight parts of water,
six to seven parts of albumen,[1] fourteen to fifteen parts
of fibrine and coloring matter,[2] some thousandths of fatty
matters, soda, salts,[3] finally traces of the peroxide of iron.
Under ordinary circumstances, those substances cannot
be discovered in the blood, which are found in the different humors formed from it in the interior of the body ; but
if the action of those organs, upon which the secretion of
those humors depends, be arrested, then we find the matters in question in the blood. It must then be concluded,
that they always exist, but in too small quantity to be
appreciated by our means of analysis, and that the organs

---

[1] *Albumen* is a matter, which enters into the composition of the majority
of the organic tissues of animals, and which forms almost of itself alone,
the white of the egg. It dissolves in water, but solidifies and becomes insoluble by heat. From the existence of the albumen in the blood, the
sugar refiners employ this liquid to clarify their syrup, as might be done by
the white of eggs.

[2] *Fibrine* forms the basis of muscular flesh. To extract it from the
blood this liquid must be beaten with rods before it coagulates ; the fibrine
attaches itself to the twigs, in the form of white, very elastic filaments.

[3] Chlorides of sodium and of potassium, phosphates, sulphates and alkaline carbonates, and carbonates of lime, magnesia and iron.

we have mentioned do not form them, but separate them
from the blood as they are developed. Wherefore the
blood may reasonably be regarded, as containing all the
materials necessary for the formation, both of the solid
and fluid parts of the body, and as well meriting the
name, conferred upon it by some authors, of *flowing flesh*.

Proportions
of the serum
and globules.
The relative proportions, in which the liquid
and solid parts, or the globules and serum enter
into the composition of the blood, vary in different animals ; and as we shall soon see, there exists a remarkable
relation between the quantity of the globules, and the
heat developed by these beings. Birds are of all animals those, whose blood is richest in globules, and those
also whose temperature is the most elevated ; the globules constitute in general fourteen or fifteen hundredths
of the total weight of this liquid. The blood of the
mammiferæ contains rather less, and in this connection a
difference exists among these animals ; in the carnivorous
and omnivorous classes the proportional quantity of globules seems to be greater than in the herbivorous ; in
man, the dog, and cat, they constitute from twelve to
thirteen hundredths of the total weight of the blood ;
while in the horse, sheep, calf, and hare, they form but
seven or nine hundredths. But the number of herbivorous and carnivorous, whose blood has been examined, is
not so great that this result can be regarded as a physiological law. Lastly, in reptiles and fishes, which are
styled animals with cold blood, on account of the little
heat they develope, the relative quantity of the globules
is much less, and hardly exceeds five or six hundredths
of the total weight of the blood.

In other respects, the proportions of the solid and

liquid elements vary in individuals of the same species, and different circumstances may occasion modifications in the blood of the same animal. The quantity of globules is greater, and that of the water less, in man than in woman, and in the blood of individuals of a sanguine temperament, than in those of a lymphatic.

It would appear, that there exists an intimate *Uses of the globules.* relation between these globules, and the vital energy; where the phenomena of life are exhibited with the most intensity, there it is found, that the blood is richest in globules, and vice versa. It is even to the presence of these particles, that this liquid owes, in a great measure, the faculty of exciting and sustaining the vital movement. The following experiment is a proof of it.

When an animal is bled to syncope, and the *Effects of hemorrhage.* flow of blood not arrested, in a few moments all muscular motion ceases, respiration is arrested, and life is no longer manifested by any exterior sign. If the animal be left in this state, the reality soon succeeds to the appearance, and death speedily arrives. But if blood, similar to that which it had lost, be injected into its veins; this *Transfusion.* apparently dead body is seen with astonishment to return to life; in proportion as new quantities of blood are introduced into its vessels, it gradually revives, and soon breathes freely, moves with facility, resumes its accustomed habits, and is completely reëstablished.

This operation, known by the name of *transfusion,* is surely one of the most remarkable ever made; and proves, better than all that can be said, the importance of the action of the globules upon the living organs; for if serum destitute of globules be employed, no more effect is produced than if cold water had been used, and

death is notwithstanding an inevitable consequence of
the hemorrhage.

But it is not as a simple physiological experiment, that
transfusion has become celebrated; it is as a curative
means, that attention has been directed to it, and its history will furnish an example of the grave errors, into
which men fall, when they would apply to practice an
incomplete science, a danger, which has caused it to be
said with some truth, that " ignorance is less dangerous
than half knowledge."

About the middle of the seventeenth century, physicians attributed almost all diseases to alterations in the
blood ; and they supposed that by changing this they
would be able to heal all ills ; thus, without having previously studied the conditions necessary for the success
of the operation of transfusion, they hastened to put it
in operation ; and Wren in England, Major in Germany,
Druis and Emmert at Paris, and several other physicians,
caused to be thrown into the veins of their patients
sometimes blood from a healthy man, and sometimes
from a calf.

Some of these attempts were not injurious, but others
occasioned the most unfortunate accidents, even death ;
and by an act of the parliament of Paris, passed in
1668, an end was happily put to these murderous experiments.

Influence of
the form of the
globules.   If, instead of prematurely applying the operation of transfusion to the art of healing, the
question had been studied in its different aspects, as has
been done of late years, these misfortunes would have
been avoided, and a means, which in some cases is really
useful, would not have been generally proscribed.   The

experiments published at London by Blundel, and at Geneva by Dumas and Prevost prove, that by proceeding in a certain manner the operation is always successful, but if an opposite course be followed, it leads to an unfortunate result. Thus the first condition of the success of transfusion, is the injection of blood flowing from an animal of the same species with that, upon which the operation is performed. If blood be introduced, differing from that of the animal in the volume of its globules, and not in form ; if, for example, the blood of a cow, or sheep, be thrown into the veins of a cat, or hare, the latter revives but imperfectly, and soon dies. Finally, if blood with circular globules be transfused into animals, whose blood contains elliptical globules, or vice versa, death takes place in a short time, and is accompanied by nervous symptoms similar to those produced by the most violent poisons.

The result then of transfusion, as it was prac- Application. tised in the seventeenth century, is not to be wondered at ; for it was inferred, that because the blood of a sheep could be introduced into the vessels of another sheep, it might also be injected into the veins of a man. But on the other hand, we see, that by employing only the blood of an animal of the same species with that on which we operate, it would not be impossible to derive a successful result from transfusion in the practice of medicine ; and in England it has been employed with marked success in several cases, where death was imminent. Notwithstanding, recourse can only be had to this operation in extreme cases, for it is always very delicate ; and if a certain quantity of air be introduced into the veins along with the blood, a thing which may easily happen, the

death of the patient is instantaneous; for this gas, having arrived at the cavities of the heart, is warmed, expands, and prevents their contraction, a mechanical obstacle, which stops the circulation.

Influence of
the blood upon
nutrition.   The influence of the blood upon nutrition is
equally easy to be demonstrated.   When the
quantity of this liquid received by any organ is diminished by mechanical means in a marked and permanent
manner, the latter is seen to diminish in size, and often
to contract, and be reduced almost to nothing.   On the
other hand, the greater the functions of any part, the
more blood it receives, and the more it increases in size.
Every one knows, that muscular exercise tends to develope the parts which are the seat of it; as in dancers, for
example, the muscles of the leg, and of the calf especially, acquire a remarkable size, while in blacksmiths,
and other men, who execute rough labor with their arms,
the muscles of the superior limbs become more fleshy
than the other parts.   Now the muscles receive more
blood when they contract than when in repose, and by
this afflux of blood the nutritive labor, of which they are
the seat, is rendered active, and their volume increases.

Influence of
the organs upon
the blood.   From the experiments we have related, it
may be seen that the blood serves not only to
repair the losses in the living organs, and nourish them,
but also to produce in certain parts an excitement, without which life could not be maintained.   Now by thus
acting upon the organs, with which it is in contact, this
liquid in its turn experiences from them modifications,
Arterial and
venous blood. and soon loses its vivifying qualities.   The
blood, on arriving in the different parts of the body, is of
a red vermilion color; while, after having traversed

them, it presents a dark tint of blackish red, and in this state does not possess the faculty of sustaining life in the organs to which it is supplied. But blood thus vitiated, or which has at least been used, retakes by the action of the air its primitive properties, and then becomes adapted to excite the vital movement.

The function, by the aid of which this important change is effected, is that of *respiration*, with which we shall soon be occupied. Blood, which has been submitted to the action of the air, and which is proper for the support of life, is called *arterial blood*; that, which has already acted upon the organs, and which cannot continue to excite the vital movement, is called *venous blood*; it contains in general fewer globules than the arterial, and coagulates less promptly, but its chief distinguishing features are, its black color, and its mode of action on the living tissues.

CIRCULATION OF THE BLOOD.

From what has been said of the part taken by the nutritive liquids in the animal economy, and of the influence exercised by respiration upon the physiological properties of these liquids, it is evident that they must be the seat of a continual movement.

Necessity of a circulatory movement.

Since it is the blood, which distributes to all parts of the body the materials necessary for their nutrition, and since this liquid is also the mean, by which the particles eliminated from the substance of the tissues are thrown out, it cannot remain in repose, and must necessarily

5

traverse constantly all the organs.   But in the majority
of animals these are not the sole conditions of existence,
which render the motion of the blood indispensable to
the support of life ; when the air does not penetrate into
the substance of all the tissues, (as is the case in insects)
but acts only by the intervention of the exterior surface
of the body, or of a special organ of respiration (as the
lungs), it is equally easy to see, that the blood, which
has already traversed the tissues, must pass to the respiratory apparatus to be submitted to the vivifying influence of the air, before returning anew to these same
tissues.

Now this actually takes place, and this movement constitutes what is called by physiologists the *Circulation of
the Blood.*

Apparatus of
the circulation.   In animals with the simplest structure, the
nutritive liquid is diffused uniformly in all parts of the
body ; it fills the spaces which the various organs, or
their constituting lamellæ, leave between them ; lastly,
it presents but slow and irregular motions.   But when
we examine beings more nearly allied to man, the blood
is seen to move in a constant direction, and there exists a
particular organ for the purpose of impressing upon it
Heart.   this motion.   This organ, called the *Heart*, is a
kind of contractile pocket, which receives this liquid in
its interior, and contracting upon itself, drives it in a determinate course.

By ascending in the scale of beings we see also, that
the blood soon ceases to circulate in simple spaces, but
moves in a system of canals, having walls proper to
themselves, and which are independent of the neighboring parts.   These canals bear the name of *blood vessels,*

and, with the heart, constitute *the apparatus of the circulation*.

The currents, of which we have just spoken,   _{Vessels.}
are seen in some animals, which have not well-formed blood vessels, and in the incubating egg they are seen before the cavities containing the blood have acquired distinct walls.   These currents may even be regarded as the determining cause of the formation of these vessels, for whenever, in consequence of certain diseases, such as fistula, a part of the body is frequently traversed by any liquid, the accidental passage thus worn is soon clothed with a membrane, and transformed into a canal, having proper walls and independent of the neighboring parts.

However, the vascular system is composed of   _{Arteries and veins.}
two orders of vessels ; of centrifugal canals, which carry the blood from the heart to the interior of all parts of the body, and of centripetal canals, which return this liquid from these organs to the heart — the former are called *arteries*, the latter *veins*.

From the functions of these vessels we can judge what ought to be their general arrangement. The arteries, having to distribute to all parts of the body the blood issuing from the heart, must necessarily be subdivided, and ramified in proportion to their distance from this organ. The veins, on the contrary, must present an inverse disposition ; they must be, at first, very numerous, and gradually unite so as to terminate at the heart by one or two great trunks. (Fig. 13.)   The arteries, as we see, may be compared to the branches of a tree, and the veins to its roots ; but they differ in one very important respect ; for in place of being separated from each other,

as the branches and the roots of plants, the arteries and
veins must be continuous, so as to form but a single system of canals, and the blood must pass from one to the
other by traversing the substance of the organs.   This

Capillary
vessels. is actually observed ; and the name of *capillary
vessels* is bestowed upon the narrow canals which unite
these two orders of ducts, and which may be considered as
being at the same time the termination of the arteries,
and the origin of the veins.

The arteries and veins, thus communicating through
one of their extremities by the intervention of the capillary vessels, are united at the opposite extremity by the
cavities of the heart ; whence it results, that the vascular
apparatus forms a complete circle, in which the blood
moves, to return unceasingly to its first point of departure, and from the nature of this movement it is called
the *circulation.*

Discovery of
the circulation. This phenomenon was unknown to the ancients ; the majority of the authors of antiquity supposed
that blood only existed in the veins, and thought, that
during life, as well as after death, the arteries were
empty, or contained merely air.   But about the middle of
the second century of the christian era, Galen proved by
delicate experiments made upon living animals the presence of this liquid in the arteries, and thus paved the
way to the discovery of the circulation.   This celebrated
man rendered to science many other important services,
and she would certainly have reaped yet greater advantage from his labors, if by a fortuitous event posterity
had not been deprived of a great part of his writings :
he left five hundred manuscript rolls, containing materials
for about eighty of our octavo volumes, and they were con-

sidered so precious, that for their better preservation they were deposited in the temple of Peace at Rome ; but this very precaution was the means of their destruction, for they were consumed together with the edifice in the reign of the emperor Commodus.[1]

In the sixteenth century some new light was thrown upon this important point in physiology. Michael Servet, known as a theologian rather than physician, and celebrated for having been burned as a heretic in a reformed city, and by the instigation of the reformer Calvin,[2] has pointed out in one of his works, the direction of the course of the blood in the pulmonary veins ; but the dis-

---

[1] Galen, one of the greatest physicians of antiquity, was born at Pergamos, a city of Asia Minor, in the year 131, the fifteenth of Adrian's reign ; he studied sometime at Alexandria, whose medical and scientific schools were then in a most flourishing condition, and at the age of 34 years he went to Rome, where he acquired by his public lectures a great celebrity, and where he excited among the other physicians so great jealousy, that he was soon obliged to quit the city, and return to Pergamos. At this moment an epidemic broke out in Italy, and his enemies profited by this circumstance, to accuse him of cowardice. He had, notwithstanding, the real advantage of enjoying during his life-time all the glory his genius could assure to him, and his high reputation remained inviolate for a long succession of ages. At the age of 38 years we find him called to Aquilea, by Marcus Aurelius, to combat a violent epidemic raging in the army of Germany, and the same prince afterwards put under his charge Commodus, his son, whose health was very delicate. Galen, however, did not delay returning to his natal city, where he died in the year 200, at the age of 69. He was one of the profoundest anatomists and physiologists of antiquity ; in this respect he may be compared to Aristotle ; and as a physician, he ranks next to Hippocrates. During a long period his reputation has been yet greater, and in the middle ages his writings were, so to speak, the sole guide of physicians.

[2] The unhappy Servet, driven by the intrigues of Calvin to fly from France, passed through Geneva where his implacable enemy was in power ; he was arrested for his religious writings, and Calvin procured his condemnation to the stake. Servet was burnt alive the 27th October, 1553.

covery of the circulation actually dates from the commencement of the seventeenth century, and the glory of
it is due to Harvey, professor of anatomy at London and
physician of the unfortunate king Charles I.   In the
lectures given by him in 1619, he pointed out the mechanism of this function ; his ideas were immediately
attacked on all sides with virulence, and when his envi-
ous contemporaries could no longer throw doubt upon
the truth of his great discovery, they attempted to wrest
the glory of it from him, by pretending that it had been
known for a long time ; they allowed Harvey only the
merit of having propagated the knowledge of it, or, as
they said, *of having circulated the circulation of the
blood :* but posterity has rendered to him full justice,
and his name will always be cited as one of the greatest
of physiologists.

Proofs of the existence of the circula-tion.   The existence of the circulatory movement
of the blood is easily demonstrated.   If we
examine by the microscope a transparent part of the
body in a living animal, the membrane which unites the
claws of the hind foot of the frog, for instance, we see
distinctly sanguineous currents traversing innumerable
capillary vessels, and continuing into other canals, yet
larger.   The direction of these currents is equally easy
to show ; if an artery be compressed, so as to interrupt
the course of the blood, this liquid is seen to accumulate
in that portion of the vessel situated towards the heart,
and to distend its walls, while beyond the compressed
point, the artery is in a short time more or less completely emptied ; it is then evident, that the blood traverses these canals from the heart to the various parts of
the body.   Now when the same experiment is made

upon a vein, the contrary effect is produced; the blood accumulates beyond the compressed point, and does not flow in the portion comprised between this point and the heart; for if the vessel be opened above and below this same point, the blood forcibly escapes from the lower, and not at all from the upper, opening. The common manner of bleeding in the arm also shows, that in the venous system the blood follows an opposite direction to what we have seen in the arteries, and returns from the various parts of the body to the heart; to swell up the vein, and facilitate the flow of blood, the vessel is compressed by means of a ligature immediately above the point where the opening is made.

In all those animals, in which respiration is _{Greater and less circulation.} made by a special organ, such as the lungs, the sanguineous vessels ramify not merely in the tissues they are to nourish, but also in the organ in which the blood must be submitted to the action of the air; and this liquid traverses two kinds of capillary vessels; one serving for nutrition, the other for respiration; the circulation carried on in the respiratory apparatus is called the *less circulation*, and that of the rest of the body the *greater circulation*.

In other respects, the course followed by the _{Direction of the blood.} blood, and the structure of the circulatory apparatus, vary much in the different classes of animals; thus, in crabs and lobsters the heart consists only of a single contractile pocket, which sends the blood to all parts of the body, whence this liquid passes into the venous system to be returned to the heart by traversing the organ of respiration. In snails, oysters, etc., the _{Crustaceous and Mollusca.} blood follows the same course, but there is a division of labor in the functions of the heart; this organ presents a more complicated structure, and is composed of a cavity called a *ventricle*, which serves to put the blood in motion, and of one or two pockets, called *auricles*, which receive this liquid from the veins, and serve as a reservoir to supply the ventricle.

Fishes.   In fishes, the structure of the circulatory apparatus is nearly the same, with this difference, that the heart, in the place of being situated at the passage of the arterial blood, belongs to that portion of the circulatory circle, traversed by the venous blood passing from the various parts of the body to the organ of respiration, which is expressed by saying, that these animals have a *pulmonary heart;* while in those spoken of above the heart is *aortic,* or belonging to the great artery of the body, which is called the aorta.

In all these animals, the entire mass of venous blood traverses the organ of respiration, and is transformed into Reptiles. arterial blood before returning to the different parts of the body ; the vessels of the greater circulation pass entirely into those of the less, and the circulation is double ; but in frogs, serpents, and other reptiles, it is more simple ; the less circulation is but a fraction of the greater, and the venous blood is not entirely changed into arterial, but mingles partly with the blood coming from the respiratory apparatus, and thus returns to the organs.

Mammiferæ and birds.   Finally, in man and all the other animals called by naturalists mammiferous, as well as in birds, the circulatory apparatus is yet more complicated. The heart presents two auricles, as well as two ventricles, and is divided into two distinct parts ; the portion situated on the left side, composed of an auricle and ventricle, corresponds to the aortic heart of snails and lobsters, and serves to send the arterial blood to all parts of the body ; while the right half of the heart, which is composed in the same manner, sends the blood to the lungs, and consequently performs the same duty as the pulmonary heart of fishes.

The blood, arriving from the different parts of the body by the venous system, first enters the right auricle ; thence passes into the ventricle of the same side, from which it is sent to the lungs by the pulmonary artery ; after having traversed the respiratory organ it returns to the heart by the pulmonary veins, which open into the left auricle ; lastly, from the left auricle the blood descends to the left ventricle, and this latter cavity sends it to the arteries, by them to be conveyed to all parts of the body, whence it returns, as we have already seen, to the right auricle of the heart.

Thus we see then, that in animals the blood passing through the circle of circulation, twice traverses the heart : in the state of venous blood on the right, and as arterial on the left side of this organ ; notwithstanding, each circulation is in itself complete, for the pulmonary cavities and the aortic cavities of the heart do not open into each other, and the entire venous blood must traverse the respiratory apparatus to be transformed into arterial blood.[1]

APPARATUS OF CIRCULATION IN MAN.

In man, which we shall take as an example   Heart.
for the illustration of the apparatus of circulation, the

---

[1] Before birth it is quite different, as will be seen when treating of reproduction.

heart is lodged in the cavity of the chest, called by anatomists *the thorax;* its inferior extremity is directed a
little to the left and forward, and its superior extremity,
whence arise all the vessels communicating with its interior, is fixed to the neighboring parts nearly in the median line of the body. In the remainder of its extent,
the heart is completely free, and it is enveloped by a
kind of double membranous sac, the *pericardium;* the
inner surface of which is in contact with itself, perfectly
smooth, and constantly moistened by an aqueous liquid,
which serves to render the motions of this organ more
easy.

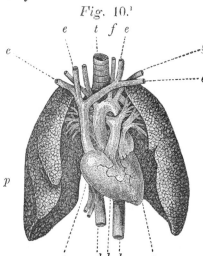

Fig. 10.[1]

The general form of
the heart is a cone, or
irregular reversed pyra-
mid; its volume is about
equal to the fist, and its
substance almost entire-
ly fleshy; it is a hollow
muscle, the interior of
which is divided by a
large vertical partition
(Fig. 11, *c*) into two
halves, each containing
two cavities, a ventricle
and an auricle. (*a, e,* and *b, d.*)

[1] *a,* Portion of the heart occupied by the left ventricle, — *b,* right ventricle, — *c,* right auricle. The left auricle is seen above the ventricle of the
same side, — *d,* vena cava inferior, — *e,* subclavian and jugular veins, ter-
minating in the superior vena cava, — *f* and *g,* carotid and subclavian arteries, originating from the arch of the aorta, — *h,* descending aorta, — *t,*
trachea, — *p,* lungs.

The two ventricles of the heart occupy its inferior portion, and their walls are endowed with a strength much greater than those of the auricles, the utility of which circumstance is evident: for the auricles have only to send the blood into the ventricles, while these latter cavities must send it to a much more considerable distance, either to the lungs,

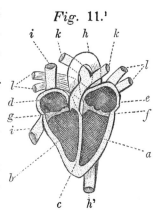

Fig. 11.[1]

or to the other parts of the body. The left ventricle is also much stronger than the right, and the comparative extent of the passage, that the contractions of these cavities must cause the blood to traverse, also explains the cause of this difference ; for the right ventricle sends this liquid only to the lungs, situated but a short distance from the heart, and the left ventricle drives it to the most remote parts of the body.

The vessels, which are to transport the arterial Aorta. blood to all the organs, spring from the left ventricle (*a*) of the heart by a single trunk, called the *aorta.* (See fig. 10 and 11, *h.*) This great artery ascends to the base of the neck, then curves downward, passes behind the heart, and descends vertically in front of the spine to the inferior part of the abdomen. In its course a great many branches are given off, the principal of which are,

---

[1] Section of the heart to display its cavities. — *a*, left ventricle, — *b*, right ventricle, — *c*, fleshy septum dividing these two cavities, — *d*, right auricle, — *e*, left auricle, — *f*, valve which separates this cavity from the left ventricle, — *g*, valve separating the right auricle and ventricle, — *h*, aorta, — *h'*, the same after its passage behind the heart, — *i, i*, venæ cavæ, — *k*, pulmonary arteries, — *l*, pulmonary veins.

the *two carotid arteries*, which ascend on the side of the
neck, and distribute the blood to the head ; the two arteries of the superior limbs, which take successively the
names of *subclavian*, *axillary*, and *brachial arteries* according as they pass under the clavicle, traverse the axilla,
or descend along the arm ; the *cœliac artery*, which goes
to the stomach, liver, and spleen ; the *mesenteric arteries*,
ramifying in the intestines ; the *renal arteries*, penetrating the kidneys ; and the *iliac arteries*, which terminate the aorta, and convey the blood to the inferior limbs.

Veins.   The *veins*, which receive the blood thus transmitted to all parts of the body, follow nearly the same
course as the arteries ; but they are larger, more numerous, and in general more superficially situated. A great
number of these vessels lie directly beneath the skin,
others accompany the arteries, and at last they all unite
to form two great trunks, which open into the right auricle of the heart, and which have received the names of
vena cava superior, and inferior. (Fig. 10, *d, e,* Fig. 11, *i.*)

Vena-porta.   The veins of the intestines present in their
course a remarkable peculiarity : the common trunk,
formed by their union, penetrates into the substance of
the liver, and there ramifies ; so that the blood of these
organs does not return to the heart, till after it has circulated in a particular system of capillary vessels contained
in the liver, and giving birth to canals, which open into
the *vena cava inferior.* This portion of the venous apparatus is called *the system of the vena porta.*

Pulmonary
artery.   The vessel, which conveys the venous blood
from the heart to the lungs, is called the *pulmonary artery* (Fig. 10 and 11, *k*) ; it arises from the superior and
left side of the right ventricle, ascends by the side of the

aorta, and soon divides into two branches, which separate almost transversely from each other, and ramify in
the lungs; that of the right side passes behind the aorta
and the vena cava superior; that of the left side passes
in front and above the arch of the aorta. The former
subdivides into three branches before penetrating into
the substance of the lungs; the second into two; they
both ramify upon the walls of the pulmonary cellules.

The *pulmonary veins* originate in the sub- ^{Pulmonary veins.}
stance of the lungs, from the minute capillary divisions
of the arteries of the same name, and unite in twigs and
branches, which follow the same course with the latter
vessels; they finally form four trunks, two from each
lung, passing to the left auricle of the heart, to which
they transmit the blood, arterialized by its contact with
the air in the interior of the respiratory organ.

The arteries and veins are lined interiorly by ^{Structure of the blood vessels.}
a thin, smooth membrane, continuous with that
lining the cavities of the heart, and analogous with those
known to anatomists as serous. In the arteries, this internal coat is surrounded by a thick sheath, yellow and
very elastic, composed of fibres of a peculiar nature, circularly disposed; and the whole is contained in a third
tunic, formed by thick close cellular tissue. In the veins,
no distinct cellular coat can be discovered, and the internal membrane is only surrounded by a thin layer of longitudinal, loose, and extensible fibres. There must be
then a very great differfence in the physical properties of
these two orders of vessels. The veins have thin flaccid
walls, which sink when they are not distended by blood,
and which quickly cicatrise, when divided. The arteries, on the contrary, have walls much thicker, and pre-

serve their calibre even when empty, as always happens after death.[1]   Lastly, when these latter vessels are opened, the edges of the wound tend to separate, on account of the elasticity of the fibres of the median tunic, and cicatrization is never completely effected, except by the obliteration of the artery at the divided point; thus to arrest the blood, which escapes from a vein, it is sufficient merely to maintain the edges of the wound for some time in contact ; but if, on the other hand, an artery has been opened, the vessel must either be tied, or obliterated by compression.

MECHANISM OF THE CIRCULATION.

Contractions of the heart.   It is easy to comprehend the mechanism, by which the blood is moved in all these vessels.   The cavities of the heart, as we have already said, contract and expand alternately, and thus drive the blood into the canals, with which they are in communication.

The two ventricles contract at the same time, and when their walls are relaxed, the auricles contract in their turn.   These contractions are called the *systole*,[2] and the opposite the *diastole*.[3]   They are very frequently renewed ; in the adult man usually from sixty to seventy-five in a minute ; in the aged their number appears to be a little augmented ; and in very young children it generally amounts to about one hundred and fifty.   Besides the age, a multitude of other circumstances influence the

[1] When death happens, the arteries contract after the left ventricle has ceased to act, so that the blood then passes into the veins and accumulates there, while the arteries remain empty ; for this reason it was so long before the uses of these latter vessels were known.

[2] Συστολὴ, from συστελλω, I contract.

[3] Διαστολὴ, from διαστελλω, I dilate.

frequency and force of the pulsations of the heart; they are accelerated by exercise, emotions of the mind, and by many diseases; in swoons and syncope they are considerably diminished, or even momentarily interrupted.

The left auricle, which receives the blood coming from the lungs, communicates, as we have seen, with the pulmonary veins on one side, and with the left ventricle on the other; by its contraction it expels from its cavity the greater part of the blood, which was there; and it is evident, that this liquid must tend to escape by two ways. This would take place, did not the ventricle dilate at the same time, and thus the greater part of the blood penetrating into its interior, very little returns into the pulmonary veins.

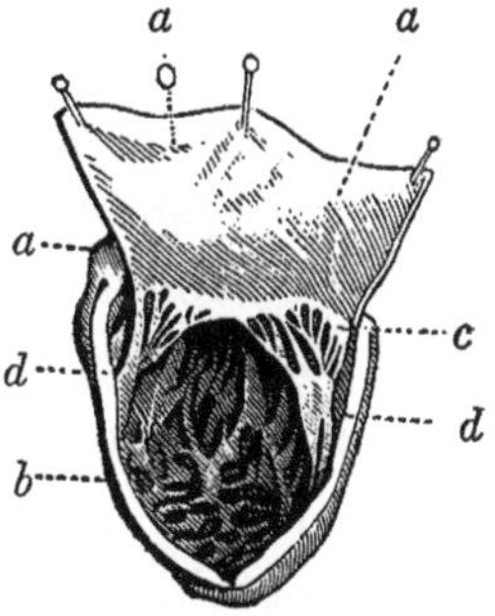

Soon after, the ventricle contracts in its turn, and drives out the blood it had just received; now, there is attached around the edges on the upper side of the opening, by which the ventricle and auricle communicate, a membranous fold, so arranged as to sink down and open when driven from above downward, and arise and shut the opening when driven in a contrary direction;[2] the result is, that during the contraction of the

Fig. 12.[1]

[1] Section of the heart to show the disposition of the valves, which separate the auricles from the ventricles. a, Auricle opened and extended. b, Cavity of the ventricle; the walls of which are supplied with a great number of fleshy fibres, irregularly disposed, so as to form a kind of cellules. c, Valve, whose external border is fixed to the circumference of the auriculoventricular opening, and the free border of which gives attachment to a great number of little tendons, (d), arising from fleshy columns fixed to the walls of the ventricle by their inferior extremity.

[2] This species of valve has been called the *mitral valve*, on account of the

ventricle, the blood cannot return to the auricle, and is driven into the *aorta*.

The measure of the force, with which the left ventricle drives the blood into the arterial system, to be sent to all parts of the body, has been the subject of much investigation, which has given the most discordant results ; thus Borelli, guided by calculation rather than direct experiment, was led to believe that this force must be sufficient for the equilibrium of 180,000 pounds (livres); while the physiologist, Kiel, estimates it at only five ounces. But a young physician, M. Poiseuille, has lately published upon this question better directed researches, and from which it would appear, that the force, with which the heart throws the blood into the aorta, is about four pounds in the human adult, and eleven pounds in the horse.

Course of the blood in the arteries. From the nature of the movements already spoken of, it would naturally be supposed, that the blood could only move in the arteries by jets, each time that the left ventricle contracted, and that, during the dilatation of this cavity, it must remain in repose. It is however quite the contrary ; if one of these vessels be opened in a living animal, the blood is seen to escape in a continuous jet, which becomes stronger at the moment of the contraction of the heart, but which is only interrupted by the opposite movement. This depends on the action of the arterial walls on the course of the

division of its free edge into two tongues. The mechanism, by means of which it closes the auriculo-ventricular opening, is very simple ; small tendinous cords, springing from fleshy columns fixed inferiorly to the walls of the ventricle, are inserted into its free border, and prevent it from turning into the auricle, while they oppose no obstacle to its opening in the opposite direction. (Fig. 12.)

blood. These walls are very elastic ; when a wave of blood is projected into the aorta by the contraction of the ventricle, they yield to the pressure thus exercised, in the manner of a spring, but they tend afterward to return upon themselves, and to drive out the blood, which distended them.

To demonstrate the influence of the arterial coats upon the course of the blood, it will be sufficient to expose a large artery in the living animal, and to intercept a portion between two ligatures tightly tied, then to make a small opening between the two points thus obliterated. The blood in this place will be completely isolated from the influence of the movements of the heart, and yet it will escape from the artery in a very elevated jet, and the vessel will soon be emptied by the simple effort of the contraction of its walls. The portion of the artery beyond the ligature diminishes in calibre, and sends into the veins the greater part of its blood.

Thus through the elasticity of the arteries, the intermitting movement impressed on the blood by the contractions of the heart, is transformed into a continuous movement. In the large arteries, the jets occasioned by these contractions may be perceived ; but in the capillary vessels, and even in the small arterial branches, they are no longer perceived, and the blood only flows through the influence of the pressure exercised by the elastic walls of the arteries.

We see then, that the contractions of the heart serve to fill continually the great arteries, and, so to speak, to stretch the spring represented by the walls of these vessels, and destined to expel in a continuous manner this liquid into the veins.

Pulse.   The phenomenon, known by the name of *pulse*, is only the movement occasioned by the pressure of the blood upon the walls of the arteries at every contraction of the heart. From the force and frequency of these movements we may judge of the manner, in which this organ beats, and draw from it useful inductions for medicine. But the pulse is not everywhere felt; to distinguish it an artery of a certain volume must be compressed between the finger and a resisting plane, a bone for instance, and we must therefore select a vessel situated near the skin, as the radial artery at the wrist.

Rapidity of the blood in the different parts of the body.   Although it be the same motive agent which drives the blood to all the parts of the arterial system, nevertheless, it is observed, that this liquid does not arrive at all the organs with the same rapidity. The distance, which separates them from the heart, is one, but not the only cause of this difference.

Influence of the curvature of the arteries.   Sometimes these vessels go in nearly a straight line, at others they form frequent elbows; now whenever the column of blood, put in motion by the contractions of the heart, meets one of these curvatures, it tends to straighten the vessel, and thus loses a portion of its motor force, which diminishes by as much the rapidity of its course.

Influence of the division of the arteries.   It is a law of physics, that, all other things being equal, the rapidity, with which a liquid flows in a system of canals, is so much the greater, as their calibre is smaller; and observation teaches us, that the total capacity of the various twigs of an arterial branch, or of the various branches of a trunk, is always superior to that of the vessels, whence they spring. It results, therefore, that the more numerous the sub-divisions of an artery before penetrating the substance of an organ, the slower must the blood be conveyed to the part; and in this respect very great differences may be observed in the animal economy; sometimes these vessels are distributed to organs only after a great number of sub-divisions, and sometimes, on the contrary, the arterial trunk buries itself in the substance of the part, in which it is to ramify.

These dispositions, by the aid of which the impetuosity of the blood is moderated at certain points of the circulatory apparatus, are principally remarked in the arteries charged with conveying this liquid to organs, whose structure is most delicate, and functions most important; the brain, for example. But nature with her enlightened foresight is not confined to such precautions, to Communication of the arteries. insure the arrival of the proper quantity of blood in each of the parts of the body. We can easily conceive, that by compression, or other accidents, an artery may be obliterated at some point of its extent, and that if the blood could not then arrive at the organ, in which this vessel is distributed, the death of the part would inevitably result, but this does not take place, for most of the arteries have frequent communications with each other, called *anastomoses*, by means of which these vessels can receive the blood of a neighboring artery, even if they do not communicate directly with the heart.

We have seen the mechanism, by which the blood passes from the heart to all parts of the body; let us now study the means employed by nature to make this liquid circulate in the veins, and to reconduct it to the heart.

Again, the contractions of the left ventricle Course of the blood in the veins. of the heart, and the obstruction of the arterial-walls, contribute most to the course of the blood in the veins.

If the passage of the blood in an artery be interrupted, and the corresponding vein be opened, this liquid will continue to flow from the latter vessel, so long as the artery by its contractions has not expelled all the blood, which distended it : but soon after, the hemorrhage will cease, although the vein be yet filled with blood, and the flow of this liquid will recommence, as soon as the circulation in the artery is reëstablished. It is then the impulse received by the blood at its departure from the heart, which causes itself to be felt in the veins, and which determines its progress in these vessels. But there are other circumstances, which tend to favor this movement.

Fig. 13.[1]

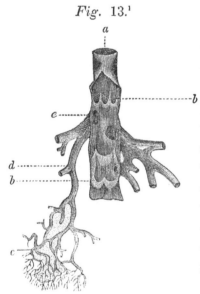

In the veins of the limbs, and of several other parts of the body, the membrane which lines these vessels, forms a great number of folds, or *valves*, which open when the blood drives them from the extremities to the heart, and close so as to intercept the passage, when this liquid is forced in a contrary direction. Now, this arrangement consequently prevents the blood from

[1] Trunk of a large vein opened to show the valves formed by the folds of its internal membrane. — *a*, Superior portion of the vein, — *b*, valves,

flowing back to the capillaries, and also contributes, in an active manner, to facilitate its passage to the heart; for, each time that the vein by the motions of the neighboring parts is compressed, the blood is driven forward, and when the compression ceases it can no longer be driven backward, but is replaced by a new quantity of liquid coming from the inferior part of the limb. All intermitting compression then of these vessels facilitates the return of the blood to the heart.

The dilatation of the chest, produced by the respiratory movements in inspiration, in the manner of a pump, facilitates also the passage of the venous blood to the cavities of the heart, as we shall see when treating of respiration.

Nevertheless, the blood flows much less quickly in the veins than in the arteries, and nature has multiplied the means proper to prevent the obstruction of one of these vessels from arresting the return of this liquid to the heart; there generally are several veins destined to fulfil the same purpose, and these vessels communicate together by numerous anastomoses.

The passage of the blood through the cavities of the right side of the heart is made in the same manner as from the left auricle to the ventricle of the same side.

Passage of the venous blood across the heart.

When the right auricle is relaxed, the blood flows into it from the two venæ cavæ; and when this cavity afterwards contracts, the greater part of this liquid passes into the ventricle, for there is around the edge of the

the concavity of which is directed towards the heart, — c, venous twigs anastomosing and uniting to form a large branch, (d), which opens into the principal trunk at e.

opening of these vessels a valve, to oppose the reflux of
blood into the inferior vena cava, and from its weight
this liquid must tend to fall into the ventricular cavity,
rather than ascend into the vena cava superior.

The opening, by which the right ventricle communicates with the auricle, is also supplied with a valve[1] similar to that of the left ventricle ; and by its contractions,
this cavity drives the blood into the pulmonary artery, by
raising other valves which surround the entrance of this
vessel, and prevent the blood from returning in the direction of the heart.[2]

Less circula-   Finally, the blood passes from the pulmonary
tion.
arteries into the veins of the same name, by traversing
the capillary vessels of the lungs, and enters the left auricle in the same manner, in which it moves in the canals
of the great circulation.

ABSORPTION.

We have seen, that the body of a living animal must
constantly assimilate to its own substance foreign materials, derived from the exterior world, and must reject
the particles separated from its own organs, and which
can no longer serve to form them.

---

[1] Called the *tricuspid valve*, because it is divided into three triangular
portions ; its arrangement is similar to that of the mitral valve.

[2] These valves, to the number of three, are formed by folds of the internal membrane of the artery, and are named, from their form, *semi-lunar
valves ;* their arrangement is analogous to that of the valves of the veins.
(Fig. 13.) When the blood is driven from the heart into the vessel, they
are raised and applied against the walls of the latter ; but when the blood
tends to reënter the auricle, the weight of the liquid distends and closes
them ; they then resemble considerably the little baskets in which pigeons

We have also seen, that a peculiar liquid, the blood, continually traverses the various parts of the economy to convey these materials.

But this nutritive liquid is itself contained in cavities within the body, which have no external opening ; the question then arises, by what way the foreign substances, necessary for the support of life, can penetrate into the vessels to be mixed with the blood, and how the materials existing in it can escape. . These two orders of phenomena constitute the functions of *absorption* and *exhalation*, with which we are now to be occupied.

Absorption is the act, by which living beings Definition. in some way suck in, and make penetrate into the mass of their humors, the substances which surround them, or are deposited in the interior of their organs.

To prove the existence of this absorbing fac- Proofs of the existence of ulty, a small number of experiments will suffice. absorption. If the body of a frog be plunged into water, so that none of this liquid can enter the mouth of the animal, nevertheless at the end of a certain time its weight is found to have been augmented ; now, this increase, which under favorable circumstances amounts to a third of the total weight of the animal, can evidently depend only upon the absorption of the water by the external surface of the body.

If a known quantity of water be introduced into the stomach of a dog, and by the aid of two ligatures all the openings, by which the cavity of this organ communicates with other parts, be closed, the liquid will yet disappear at the end of a little while, for it will be absorbed

---

are hatched ; and as they touch at their free edges they close the artery. Similar valves exist at the entrance of the aorta, where they serve to prevent the blood from reëntering the left ventricle, while this cavity dilates.

by the walls of the stomach, and thus mingled with the blood.

Mechanism of absorption. And yet there do not exist on the surface of the skin, or stomach, any pores[1] or openings conducting directly to the blood-vessels, and which may serve for a passage to the liquids absorbed. But the tissues, which form these organs, as well as those of all other parts of the body, have a structure somewhat spongy, and are more or less permeable to liquids.

Imbibition. In the living, as well as the dead body, these tissues always imbibe the fluids, which bathe them, and allow themselves to be traversed with more or less facility.

Fig. 14.[2]

b   *a*

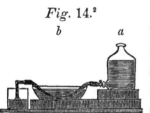

Thus, if a current of acidulated water be made to traverse a section of a vein, arranged as in the figure, the external surface of this vessel being in contact with a solution of tournesol, the color of the latter liquid will soon be changed to red by the action of the acid, which has passed through the walls of the vein. In the dead body consequently these parts are permeable to liquids.

Now, if we expose a vein in the living animal, perfectly isolate this vessel, and apply upon its exterior surface the extract of nux vomica, this violent poison will

[1] The pores, which are perceived upon the surface of the skin, do not traverse this membrane, but only conduct to small cavities lodged in its interior, and serving to secrete various humors, or to form the hair; when treating of the touch we shall have occasion to return to the structure of the skin.

[1] *a*, Flask with two openings containing the acidulated water, and serving as a reservoir; *b*, vase containing the blue solution of tournesol into which is plunged the median portion of a vein, one extremity of which communicates with the reservoir, *a*, and the other with the vase, *c*, which receives the acidulated water flowing through the vein.

soon penetrate the membranous walls of the vein, mingle with the blood, and occasion the terrible symptoms, which are observed, whenever it is directly injected into the blood-vessels. It is then evident, that, during life as well as after death the veins are permeable to liquids.

The permeability of the solid parts of organized bodies will at once explain to us in what manner absorption is possible. By the aid of this property of the living tissues, the liquids may have access every where ; but this would not be enough, and, that they may penetrate into the interior of the organs, they must be attracted by some force.

The capillary action contributes powerfully to Capillarity. produce this imbibition ; but it is not the only force which acts in this way ; and to form an exact idea of the mechanism, by means of which liquids penetrate the substance of the organic tissues, we must understand a very curious phenomenon, recently discovered by M. Dutrochet, and named by him *endosmosis.*

This physiologist has proved, that, if a Endosmosis. solution of gum be inclosed in a small membranous sac surmounted by a tube, and surrounded by pure water, the latter liquid will penetrate into the interior of the apparatus, and ascend in the tube to a considerable height. Here then there is a true absorption, and the force, which determines it, acts often with such energy as to constitute the equilibrium to a column of water of several centimetres. If, on the contrary, gum or sugar-water be placed without the membranous sac, and pure water in its in-

Fig. 15.

8

terior, the passage takes place inversely, and the sac, instead of filling, is emptied.

This phenomenon is very analogous to the absorption which takes place in living beings, and the explanation is easily given. We have seen, that the organic membranes, as well as all the spongy and porous bodies, may be traversed by liquids ; but the facility of this transport varies with the fluidity of these liquids, and the ease with which they imbibe the filtrations. If the two liquids placed in the interior, and upon the exterior of the membranous sac, could traverse equally well the walls of this cavity, they would mingle, and the same level be established upon either side. But if the exterior liquid traverse more readily the walls of the sac than the interior, the current from without inward will be more rapid than the current in a contrary direction, and the liquid will accumulate in the interior of the apparatus. Now, this takes place in endosmosis ; the water surrounding the sac which contains the gum-water, filtrates easily through the walls of this cavity, and when it has reached the interior, it unites with the gum, and thus forms a new liquid, the passage of which through these walls is difficult in proportion to the quantity of gum. It must then accumulate, and ascend in the vertical tube, which communicates with the membranous reservoir.

The organized bodies, which absorb the liquids by which they are surrounded, are placed in the same conditions with the membranous sac, of which we have now spoken ; the presumption then is, that in all these cases the same effects arise from analogous causes, and that the principal force, which occasions the passage of the absorbed substances across the living membranes, is the same with that, which produces the phenomenon of endosmosis.

In certain animals of the inferior classes, those with the least complicated structure and most limited functions, absorption consists only in the kind of imbibition, of which I have spoken. By the same mechanism foreign substances traverse the solid parts, with which they are in contact, to be mingled with the liquids, by which the areolæ of these organs are filled, and are afterwards diffused through all the body and penetrate the interior of the tissues. But as we ascend in the series of beings we soon see that nature perfects the mechanism of absorption, and that to accomplish this, she introduces into this important function a constantly progressive division of labor.

In the animals possessing a regular circulation, the absorption properly so called, or the passage of foreign substances from without into the interior of the economy, is always effected in the same manner as in beings less perfect; but from the moment when these substances are mingled with the nutritive juices of the body, a great change takes place; instead of being gradually diffused to the various parts by the effect of imbibition, they are taken up by currents more or less rapid, and immediately distributed to all the points, to which the blood penetrates. We see then, that the absorption of these materials, and their transport to the interior of the economy, are no longer a single act, but are composed of two series of phenomena perfectly distinct; the first, purely local, consists in the imbibition of the tissues, and in the mingling of the matters absorbed with the humors of these parts; the second, dependent upon the general circulation, consists in the transport of these same substances to parts remote from those, at which they originally penetrated.

Venous absorption.   In all these beings the principal agent by which this transport is effected, is the blood, which traverses the organs in which absorption takes place, and which returns again to the heart, to be afterwards conveyed anew to the interior of the various tissues. It follows that, in animals provided with a circulatory system, the veins perform a very important office in the absorption ; and that, in the immense majority of cases, it is by their intervention, that the liquids, by which a circumscribed part of the body is surrounded, are diffused through the whole economy.

Lymphatic absorption.   In very many animals, absorption is only effected through the blood-vessels ; but in man, and most animals with the more complicated organization, there exists another system of canals, which answer the same purpose, and which serve especially to absorb certain substances. This is the apparatus of the *lymphatic vessels*.

This name is given to canals, which spring from very minute radicles in the interior of the different organs, but afterward unite in trunks of varying size, and finally empty into the veins near the heart. Many physiologists regard these canals as the sole agents of absorption, and call them *absorbent vessels ;* but this opinion is without foundation. Comparative anatomy alone would overthrow it, and the experiments of Magendie and many others prove it to be completely erroneous.

Proofs of the Venous absorption.   The absorption by the veins is as easily proved in all animals with a system of lymphatic vessels, as in those without. The following experiments can leave no doubt in this respect.

Messrs. Magendie and Delille, having stupefied a dog with opium, to render him insensible to the pain occasioned by a laborious operation, amputated one of his thighs, leaving only the artery and vein untouched, to maintain the communication between the limb and the rest of the body ; then they deposited within the paw thus separated a violent poison (*Upas Tieuté*). The effects of the poison were manifested with the same promptitude and intensity, as if the limb had not been severed from the body ; and the animal perished in a few moments.

It might be objected that, notwithstanding the precautions taken, the untouched coats of the artery and vein contained in their tissues the lymphatic vessels, and these canals were sufficient to give passage to the poison.

To remove this difficulty M. Magendie repeated the experiment upon another dog, with this modification, that he introduced into the femoral artery a quill, to which he fastened the vessel by two ligatures, he then divided circularly between the two the walls of the artery, and performed the same operation upon the vein. The only means of communication then, between the thigh of the animal and the remainder of his body, were, the arterial blood coming to the limb, and the venous blood returning to the heart ; yet the poison then introduced into the paw produced death with its ordinary rapidity.

This experiment leaves no doubt, but that the poison passes from the paw to the trunk through the crural vein ; and to render the proof yet more conclusive, it will be sufficient to press this vein between the fingers at the

moment when the effects of the poison begin to be manifested; for by thus impeding the passage of the blood,
the symptoms of poisoning cease at once, to reappear as
soon as the vessel is again left free, and the blood allowed to ascend to the heart.

In other experiments, the presence of the materials
absorbed has been directly proved in the blood of the
veins.   It is then evident, that these vessels are active
organs of absorption, but it cannot be doubted, that the
Proofs of the lymphatics in certain cases discharge the same
lymphatic ab-
sorption.   duties.   As we shall hereafter see, these latter
ducts are especially charged with the transport of the nutritive materials, extracted from the aliments by the labor
of digestion, and in the other parts of the body they appear to fulfil analogous functions.   M. Dupuytren, in
making the autopsy of a patient, who sank under an enormous abcess of the thigh, found the neighboring lymphatic vessels distended by a liquid, having all the characters of pus.   And it has for a long time been known,
that, if when dissecting a putrified body an individual
prick his finger, there often arise grave accidents from
the absorption of the substances thus inoculated, and it
is not rare to see the lymphatic vessels, which extend
from the wound to the trunk, swollen, and inflamed, as
if the passage of the poison had irritated their walls.

Moreover the absorption by the lymphatics must be
slower than that by the veins; for the blood flows with
great rapidity in these latter canals, and the liquid contained in the lymphatic vessels moves but slowly.

Lymph.   This liquid is called the *lymph*.   Its physical
properties are not always the same; sometimes it is opaline, and of a faint rosy tint; at others yellowish, and

sometimes red ; examined by the microscope, a multitude of small globules are seen in it analogous to those of the blood, and when left to itself it coagulates like the latter, and separates into two parts, a serous fluid, and a solid clot, which exposed to the action of the air takes a red tint.[1]

The lymphatic vessels resemble the veins in Lymphatic vessels. their structure and mode of distribution, but they are much finer, and their walls thinner. They are met with in nearly all parts of the body ; they form in general two planes, one superficial, the other deep seated ; they communicate by frequent anastomoses, and unite in twigs and branches like the veins. The majority of these vessels thus form one large trunk (called the *thoracic canal*), which ascends in front of the vertebral column, and empties into the subclavian vein of the left side ; but others open separately into the vein on the opposite side of the neck, or even sometimes into the different blood-vessels, situated nearer their origin. During their course they are seen to traverse little organs, irregularly rounded, and situated in the axillæ, groin, neck, chest and abdomen. (See Fig. 25.) The structure and uses of these bodies are yet but little known ; they are called *ganglions*, or *lymphatic glands*. Finally, in the interior of the lymphatic vessels there exist a great number of transverse folds, which discharge the same functions as the valves of the veins, and which oppose the reflux of the lymph.

From what has already been said upon the Circumstances influencing absorption. mechanism of absorption, it will readily be per-

[1] By chemical analysis the clot of the lymph has been found to be composed as follows, viz.: water, 925 ; fibrine, 3 ; albumen, 57 ; alkaline, chlorides, soda and phosphate of lime, 14 ; to the thousand parts. The proportion of the clot to the serum appears to be nearly as 1 to 300.

ceived, what the principal circumstances are which influence the course of this function.

Permeability and vascularity of the tissues. Thus, the first condition of all absorption being the permeability of the tissues interposed
between the substance to be absorbed, and the liquids
which are to effect its transport; it is evident, that, other
things being equal, this phenomenon will be rapid in
proportion to the lax and spongy texture of this tissue itself, and the degree of vascularity, which is the seat of it.

In truth, the lax and spongy texture of the organic
solids is of all the physical properties, that which most
facilitates imbibition; and with regard to the veins, since
by them principally the absorbed substances are diffused
in the economy, the influence of their number and size
is too evident to require any comments. In the majority of cases, these two laws will explain to us the great
diversity, observed in the rapidity, with which absorption
is effected in the various parts of the body; we might
even anticipate this difference from the sole consideration
of the anatomical disposition of our organs.

Thus the lungs, the structure and functions of which
will be examined hereafter, are of all parts of the economy those, in which the structure is most spongy, and the
vascular system most developed. It follows, that absorption must be more rapid in these organs than elsewhere,
and to this result does experiment lead.

The soft and whitish substance, which is found between all the organs, and which is called the *cellular tissue*, is also very permeable to liquids, but possesses far
fewer blood-vessels than the tissue of the lungs; therefore absorption, although very rapid, takes place less rapidly than in these organs.

The skin presents, on the contrary, a very dense texture, and its surface is covered by a kind of varnish, formed by the epidermis; in general, its blood-vessels are small and few; and, as might be expected from this anatomical disposition, absorption takes place with great difficulty. By raising the epidermis, imbibition is considerably facilitated, and consequently absorption is rendered more easy; finally, when we do not confine ourselves to simply stripping the dermis, but also excite its vascular system, (by the irritation of a blister, for example) this function is rendered yet more active.

In medicine, use is made of this fact to obtain the absorption of certain substances, whose irritating nature upon the stomach is feared, and this mode of administering remedies is called the *endermic method.* The slight degree of permeability of the epidermis also explains to us how the most violent poisons may be made use of without danger, provided the skin of the hands be untouched; for in them absorption is nearly null; while the gravest accidents, and death itself, may result from the contact of these same substances at any point, where the skin has been wounded, or merely deprived of the epidermis.

Another circumstance, which also exercises a great influence upon the rapidity of absorption, is the state of *plethora* of the animal.[1] Mass of the humors.

The quantity of liquid, which may be contained in the body of a living animal, has limits as well as the degree of desiccation, compatible with life. Now, the nearer the body approaches its point of saturation, the

---

[1] The word *plethora* (πληθώρα, πλήθω, I fill) is employed to indicate the state of plenitude of the vascular system.

greater difficulty do the liquids experience in reaching its interior.

Thus, if to two dogs equal doses of a poison be administered, the effects of which are not manifested till after absorption; and if previous to this operation, the mass of humors in one of these animals be diminished by a copious bleeding, while in the other, the volume of liquids contained in the body be increased by the injection of a certain quantity of water into the veins; the effects of the poison will take place sooner in the former than in ordinary cases, and in the latter those symptoms, which denote the absorption of the poison, will not appear for a much longer time.

These results are more important to be known, as they meet with constant application in the healing art, and show how much the functions of living beings are subject to the ordinary laws of physics. The researches of my brother, Dr. W. Edwards, relative to the influence of physical agents upon life, have fully established this truth, and M. Magendie has arrived at the same result by following a different course.

Nature of the absorbed substances.   Lastly, the nature of the substances absorbed also exerts an influence upon the ease, with which they penetrate into the interior of the tissues, and are carried into the current of the circulation. In general thesis we may say, that cœteris paribus, the absorption will be so much the more rapid, as the liquids are less dense, and readily moisten the tissues. In order that a solid may be absorbed, we must consider in the first place its degree of solubility, and next, the physical properties of the solutions it forms.

Thus, if water be injected into the abdominal cavity

of a living animal, this liquid will promptly disappear; while oil, placed in the same condition, is not sensibly diminished in volume for a considerable lapse of time.

———

Such are the most important points in the history of absorption. Let us now study the inverse function, that, by which a portion of the substances contained in the general mass of the humors, and enclosed with them in the blood vessels, can escape, either to penetrate the cavities within the body, or to make its way externally.

———

EXHALATION AND THE SECRETIONS.

The passage of the fluids from the interior of the vessels outward, may take place in three different ways; sometimes a portion of the blood itself is expelled from these canals with all its constituent parts, which is called a *sanguineous effusion;* in others, merely a portion of the aqueous part of the blood issues from the vessels, carrying with it a certain quantity of the soluble materials, already existing in the liquid, and this phenomenon is styled *exhalation;* lastly, there are also times, when there are separated from the blood new products, which differ from it by their acidity, or greater alkaline properties, and which often contain in abundance substances, of which but the slightest trace exists in the blood. This labor, in some sort chemical, constitutes what is called by physiologists a *secretion.*

TRANSUDATION, OR SANGUINEOUS EFFUSION.

The mechanism, by the aid of which the blood is effused from the vessels to escape externally, or to flow into the cavities within the body, is very simple. In some parts of the economy the veins communicate with a peculiar spongy tissue by means of openings, which in the state of repose are not open, but which afford a free passage to the blood, if any obstacle whatsoever, by opposing its ordinary course, determines its accumulation in the veins. By such a process does the *erectile tissue*, which forms various appendages around the head of turkeys, for example, swell, extend, and take an intense red color.

At other times, the sanguineous effusion takes place without any such perceptible openings to the veins; and then this liquid appears to filtrate through the substance of the tissues; this is observed but in a small number of cases, and generally arises from a pathological condition.

EXHALATION.

Exhalation is equally a physical phenomenon, the course of which may be modified by the action of the vital forces, but its existence is independent of them.

Mechanism of the exhalation. We have already seen, that the walls of the blood vessels, as well as the other parts of the body, are permeable to liquids; now we can easily comprehend, that the most fluid part of the blood must traverse them much more readily than the solid corpuscles contained in this liquid, and that, by acting as a filter, these membranes must produce the phenomenon of exhalation.

This actually takes place, both in the dead and living body ; if there be thrown into the arteries of a dead animal a solution of gelatine, colored by vermilion reduced to a very fine powder, the injection will penetrate the capillary vessels, and then we often see a portion of water traversing their walls loaded with gelatine, to escape externally, while the coloring matter is retained in their interior.

The mechanism of exhalation is the same with that of absorption ; all the parts, which are the seat of one of these functions, may be of the other : in general they take place simultaneously, and every thing, which tends to modify the course of one, influences the other.

The degree, to which the texture is spongy, Circumstances which influence exhalation. and of course more or less favorable to imbibition, is a condition, acting equally upon absorption and exhalation. These functions, cœteris paribus, are active in proportion to the number of blood vessels traversing their seat.

The variations in the mass of the liquids contained in the body act, on the contrary, in an inverse manner upon these two functions, the greater the quantity of liquids, the more abundant the exhalation. In the living body as in the dead, the tissues retain water more powerfully, the less they contain, and by increasing the mass of humors we can at will make active the exhalation.

Finally, the pressure, which the blood supports in the vessels, also exerts a powerful influence upon the exhalation ; and when the circulation in the veins is interrupted, so as to cause the accumulation of this liquid, the more fluid portion of the blood exhales in abundance into the neighboring parts, and causes them to swell, which produces the tumidity of those parts, which have been closely surrounded by ligatures.

External and internal exhalations.   Exhalations are divided into external and internal, depending upon the fact, whether they take place on the general surface of the body, or in cavities, which have no free external communication.

The *exterior exhalation*, which must not be confounded with the perspiration, and which takes place on the pulmonary surface, as well as upon the skin, is also styled *insensible transpiration*, because its products are dissipated by evaporation, and are not usually appreciable to our senses. The losses experienced by man and other animals in this way are very considerable. In the state of health, the weight of the body of an adult hardly varies, and the losses, which he experiences by the various excretions, counterbalance the weight of the aliments daily used by him ; now from the experiments of Sanctorius it appears, that insensible transpiration often constitutes five eighths of the total losses, of which we have spoken.

However, the evaporation, going on upon the surface of the body, does not always take place with the same intensity, and even here the influence of physical agents is felt in nearly the same manner upon the living and the dead. In both, the losses by evaporation are augmented by the elevation of the temperature, agitation of the air, (winds, etc.) by its dryness, by the diminution of the atmospheric pressure, etc.

The *internal exhalations* take place upon the surface of the walls of cavities, varying in size, situated in the interior of the body, and they also consist of water, mingled with a small quantity of animal matter, and salts contained in the blood, whence these liquids escape.

Such is the source not merely of the humors which continually moisten the serous membranes,[1] by which the
great viscera of the head, chest, and abdomen are enveloped ; but also of the serosity, which bathes the lamellæ
of the cellular tissue, so abundantly diffused in all parts
of the body ; and of a part of the humors, which fill the
interior of the eye.

As these internal exhalations take place upon the surface of cavities, which have no outlet, it is evident, that
the quantity of liquids contained in this species of reservoirs would constantly augment, if the parts, thus exhal-
ing, were not at the same time the seat of an absorption
not less rapid.   In the state of health, these two functions are exercised simultaneously, and counterbalance,
so as to maintain always the same quantity of liquid in
the interior of the cavity ; but sometimes this equilibrium
is destroyed, and exhalation becomes more active than
absorption ; the liquids then accumulate in the parts, and
diseases result known as Dropsies.[2]

[1] The disposition of the *serous* membranes requires attention ; they have
always the form of a species of sac, whose internal surface extremely
smooth and moistened constantly by a liquid, is everywhere in contact with
itself ; one of the halves of this sac adheres by its external face to the walls
of the cavity lodging the viscera, and the other half surrounds these viscera
and adheres to them by its external face.   To make use of a trifling comparison, but one which perfectly represents the thing, these membranes
resemble a double night-cap, and surround the viscera as this cap the head,
the exterior surface of which should be fixed to the walls of a cavity containing both the cap and the head.   These membranes serve to diminish
the friction of these parts, and consequently to facilitate their movements ;
therefore similar sacs are met with wherever the organs continually, or
forcibly, rub against each other, as at the articulations of the bones of the
limbs, around the intestines, etc.

[2] These collections of water are variously denominated according to the
parts, which are the seat of them ; the special name of *dropsy* (or *ascites*) is

SECRETIONS.

The *secretions* differ essentially from the exhalations, in that the liquid separated from the blood is not merely water, or serum, but a humor, the chemical nature of which is wholly distinct from that of the blood, or its serum.

Theory of the secretions. The blood, as we have already seen, is slightly alkaline ; the liquors secreted are sometimes very alkaline, at others acid ; and in them we find particular substances, not existing in the blood, or in quantities too small to be appreciated by our means of analysis. Here then chemical action takes place, and by comparing the phenomena of the secretions with those produced by the action of the voltaic pile, a striking analogy may be observed. If an electric current be passed through a liquid, holding in solution salts and albumen, serum for example, there is formed at one pole of the pile an acid, and at the other an alkaline liquid, and animal substances dissolved in it are seen to change their nature. Now, it is precisely the same with the secretory organs ; and by admitting, that the one are seat of the positive, and the others of the negative pole of an electric apparatus, the greater part of the phenomena met with would be accounted for very easily. But this theory, plausible as it is, cannot be received until based upon facts, and unfortunately these facts are wanting.

given to accumulations of water in the cavity of the abdomen ; and *hydro-thorax*, or *dropsy of the chest*, to such as are formed in the pleura, the membrane surrounding the lungs ; *dropsy of the heart*, such effusions as take place into the pericardium, the membrane around the heart ; *hydrocephalus*, those in the membranes covering the brain ; and *œdema*, those exhibited in the cellular tissue of the various parts of the body.

However, the secretions are not made indif- ferently in all parts of the body as the exhalations; they always have their seat in special organs, which have a very peculiar mode of structure. They are always composed of a greater or smaller number of extremely minute cavities, in the form of pockets, purses, or canals, of very great tenacity, and which receive a great number of blood-vessels, as well as of nerves. They are designated by the general name of *glands*, and divided into *perfect* and *imperfect*, according as they are furnished with a duct through which the product of their secretion is poured out, or as they have the form of cavities without openings, and from which the liquids secreted can only issue by absorption.

The disposition of the *perfect glands* greatly varies; some are situated near the surface of various membranes, and open directly into them, without having an excretory canal in the form of a tube : these are called *simple glands*, or *crypts*. Others consist of masses of crypts, which empty the products of their secretions by several openings; these are the *agglutinated glands*. Lastly, others still present excretory ducts in the form of ramified tubes, and which unite into a small number of canals ; they bear the name of *conglomerated glands*, and in the course of their excretory duct, there is sometimes found a membranous pocket, serving as a reservoir for the liquid secreted.

As an example of the crypts we will cite the *follicles* scattered upon the mucous membrane of the digestive canal, also those which open upon the surface of the skin, and secrete the fatty and unctuous matter, by which the hair is moistened ; the tonsils belong to the class of

10

the agglutinated glands ; and the liver, kidneys, salivary glands, etc. to the conglomerated glands.

The *imperfect glands* are formed by little pouches disseminated in the cellular tissue, or collected in masses of varying volume. The organs, which secrete the fat, and which are lodged in the interior of the cellular tissue, present the former of these dispositions ; in very lean persons it is difficult to distinguish them, and they are confounded with the cellular tissue ; but when filled with fat, they are seen to be formed of a very thin membrane, rounded in form, without opening.[1]

---

[1] The *fat* is essentially composed of two particular materials, *elaïne* and *stearine*, one liquid and the other solid at the ordinary temperature ; the relative proportions of these two substances vary greatly in different animals, and hence corresponding differences in the consistence of their fat. In general, the principal uses of this matter are all mechanical, and it serves as an elastic cushion to protect the organs it surrounds ; this is seen in the orbit, where the eye reposes upon a thick bed of fat, on the sole of the foot, where it is also found in considerable quantity, and in other parts of the body exposed to constant pressure, or friction. It may also, from its feeble power of conducting caloric, contribute to preserve the heat disengaged in the interior of the body ; finally, it may be considered as a kind of reserve of nutritive materials deposited in certain parts of the body, in order to serve the purpose of assimilation, when the animal can no longer derive from without the substances necessary for the maintenance of life ; when fat persons remain a long time without eating, their fat is gradually absorbed, and appears to serve for their nutrition ; it is also remarked, that the hibernating animals, which pass a great part of the cold season in a state of lethargy, are surcharged with fat when they become stupid, and are on the contrary very lean when they awaken from their sleep of many months. Fat is not deposited with the same facility in all parts of the body ; it abounds especially in the folds of the mesentery, (a portion of the peritoneum which envelopes the intestines,) around the kidneys, and under the skin. Repose exercises a great influence upon its formation ; very young children are usually very fat, but when they begin to take much exercise, their fat is gradually dissipated, and while the body rapidly increases it is rarely deposited in considerable quantities.

Among the imperfect and massive glands we will cite the thyroid body,[1] and the thymus,[2] organs, whose uses are not yet known.

The liquors produced by the secretions are, _{Humors secreted.} as we have said, acid or alkaline. The most important alkaline humors are the bile, formed by the liver ; the saliva, produced by the salivary glands ; and the tears, secreted by the lachrymal glands. The principal acid humors are the urine, elaborated by the kidneys ; the perspiration, which distils from the follicles of the skin ; the mucus, which lubricates the mucous membranes, and issues from the crypts so abundant upon their surfaces ; and the milk, which is secreted by the mammary glands. Hereafter we shall be obliged to return to the study of these liquids, and to point out their properties and uses.

RESPIRATION.

Having investigated the manner, in which circulation, absorption, and exhalation take place, we may turn our

---

[1] The *thyroid body* is an ovoid mass, soft, spongy, and glandular in appearance, situated at the anterior and inferior part of the neck, in front of the trachea. It is in general larger in the infant than the adult, and exists in all the mammiferæ, but is wanting in birds, most reptiles, fishes, and other animals of the inferior classes. The swelling of this body constitutes the tumor called *goître*.

[2] The *thymus* is a glandiform mass enclosed within the chest between the two folds of the anterior mediastinum, (a cavity formed by the union of the exterior surfaces of the pleuræ, and which lodges the heart.) It is extremely developed in the fœtus ; but soon after birth its volume is much diminished, and in the adult it is completely atrophied.

attention to the study of another function, the history of which is closely united with that of the blood, and in importance not inferior to it : I mean the *respiration*.

We have seen, that the arterial blood, by its action upon the living tissues, loses the qualities, which render it proper for the maintenance of life, and that after being thus modified, it resumes, by contact with the air, its former properties ; this contact is then necessary to the existence of living beings.   And, if an animal be placed under the receiver of an air pump, in which a vacuum is created, or if it be deprived of air by any other means, a great trouble at once arises in the various functions ; the action of all the organs is soon interrupted, life ceases to be manifested, and the animal falls into a state of asphyxia, or apparent death ; at last, life becomes entirely extinct, and can no longer be recalled.

This phenomenon is one of the most universal in organic nature ; the contact of air is indispensable to all animals as well as vegetables, and a living being deprived of it always dies.   Wherever there is life, air is necessary.

At first one would suppose, that the animals living at the bottom of the water, as fishes, were withdrawn from the influence of the air, and consequently exceptions to the law just given ; but it is not so, for the liquid, in which they are plunged, absorbs and holds in solution a certain quantity of air, which they can easily separate, and which suffices for the support of life.   It is impossible for them to exist in water destitute of air, for they are asphyxiated, and die in the same manner as mammiferæ, or birds, when withdrawn from the action of the atmospheric air under its ordinary form.

The relations of the air with organized beings form one of the most important parts of their physiological history, and the series of phenomena, which result from it, constitute the act of RESPIRATION.

Air, we say, is necessary to the life of all animals, but this fluid is not a homogeneous body; chemistry has demonstrated in it the existence of very different elements, and which consequently cannot take the same office in the performance of respiration. Besides the watery vapor, with which the atmosphere is always more or less loaded, the air furnishes by analysis twenty-one hundredths of oxygen, and sixty-nine hundredths of azote, as well as traces of carbonic acid gas. The first question, which presents itself to the mind when we enter upon the study of respiration, is to know, whether these different gasses act in the same manner, or if to one in particular belongs the property of sustaining life.

The vivifying properties of the air depend upon the oxygen contained in it.

To ascertain this a small number of experiments will be sufficient. If a living animal be placed in a vase, filled with air, and all communication of this fluid with the atmosphere be intercepted, at the end of a longer or shorter time the animal will become asphyxiated, and perish. The air, which surrounds it, has then lost the faculty of sustaining life, and by chemical analysis it is found to have lost at the same time the greater part of its oxygen. If another animal be placed in a jar, filled with azote, it also perishes; while a third animal put into oxygen breathes with more activity than in the air, and presents no symptom of asphyxia. It is evident therefore, that the atmospheric air owes its vivifying properties to the presence of oxygen.

The discovery of this important fact dates only from
the close of the last century (1777), and is due to one
of the most celebrated French chemists, Lavoisier, who,
notwithstanding his numerous titles to public favor, perished prematurely a victim of the revolution.

Production of carbonic acid. By the act of respiration, we have said, all
animals remove from the air, which surrounds them, a
certain quantity of oxygen ; but the changes, they thus
determine in the composition of this fluid, are not so
limited ; the oxygen, which disappears, is replaced by a
new gas, carbonic acid. The production of this substance is not less universal among animals, than the absorption of oxygen ; and in these two phenomena essentially consist the performance of respiration.

To prove this fact, we have only to blow, during a
certain time, through a tube into water holding lime in
solution. The carbonic acid has the property of uniting
with this latter substance, and thus giving origin to a
body, which is insoluble, and which in its composition is
analogous to chalk ; now, in this experiment, the carbonic acid, which escapes from our lungs, combines with
the lime, and forms a whitish powder, which by its deposition troubles the water, and is easily perceived. By
such means, in 1757 a chemist, named Black, first proved
the production of this gas during respiration. Carbonic
acid may also be recognised by other methods, for it
extinguishes bodies in combustion, and causes the destruction of animals when inspired even in small quantities.[1]

[1] Carbonic acid, which is formed by carbon united in certain proportions
with oxygen, is produced by the combustion of charcoal, during the alco-
holic fermentation, &c., — it enters into the composition of marble, chalk,

As to the azote of the respired air, its volume   _{Azote.}
changes but little, and its principal use appears to be to
weaken the action of the oxygen, which in the state of
purity excites animals too strongly, and produces in them
a kind of fever.

It has been remarked, however, that in some cases, a
part of the azote of the air disappeared during respiration, and in others, its volume was augmented. It would
even appear, that animals absorb and exhale it continually, as they exhale and absorb the liquids contained in
the cavity of the pericardium, peritoneum, etc., and that
the variations observed depend upon this ; that these two
opposite functions are in general in equilibrium, so that
their result is not apparent, but that the absorption being
sometimes more active than the exhalation of the azote,
while at other times the quantity exhaled exceeds that
absorbed, occasions a diminution, or increase in its vol-

&c., and is found in most mineral waters. In the state of gas it is colorless as air, but much heavier than this fluid, and soluble in water. From
the action of this acid upon the animal economy arises the asphyxia, produced by the vapor of charcoal, as well as most of those accidents, which
take place in mines, drains, wells, and vaults where wine or beer is fer-
menting. In a vault near Naples it is continually disengaged from the
interior of the earth, and occasions phenomena, which at first seem very
singular, and excite the curiosity of all travellers: when a man enters this
cavern he experiences no difficulty in respiration, but if accompanied by a
dog, the animal soon falls asphyxiated at his feet, and would quickly perish,
if not taken into the fresh air. This depends upon the fact, that the carbonic acid, being heavier than the air, does not ascend, but remains near
the ground, and there forms a layer about two feet thick. Now a dog entering this grotto is necessarily entirely plunged into this mephitic gas, and
must be asphyxiated ; while a man, whose height is more elevated, has
only the lower part of his body exposed to the action of the carbonic acid,
and breathes freely the pure air above. This remarkable place is called the
Grotto of the Dog.

ume when compared before, or after, it has served for respiration.

Pulmonary transpiration. Lastly, there also escapes from the body with the products of respiration a considerable quantity of watery vapor. This exhalation, which has received the name of *pulmonary transpiration*, is one of the most apparent phenomena of respiration, when by the refrigerating action of the surrounding air these vapors are condensed on their departure from the body, and form a thick cloud.

Modifications of the blood. While the respired air undergoes the changes already indicated, the blood, which passes over the membranes in contact with this fluid, also experiences important modifications; it is rendered proper for the support of life, and changes from a blackish color to a lively and sparkling red. To observe this fact, we have only to open an artery in the living animal, and to compress at the same time its neck, so as to prevent the air from penetrating into the lungs; the blood flowing from the artery will at first be of a lively red, but will soon become dark, and of a venous color. If a new access of air to the lungs be then permitted, this liquid is seen to change its color, and take the tint proper to the arterial blood.

Theory of respiration. Such are the principal phenomena of the respiration of animals. Let us now seek for an explanation of them.

And first, what becomes of the oxygen, which disappears, and what is the origin of the carbonic acid produced during the exercise of this function?

When charcoal is burnt in a vase filled with air, oxygen is found to disappear, and its place to be supplied by an equal volume of carbonic acid gas ; at the same time a considerable disengagement of heat takes place. Now, during respiration, the same phenomena take place, and a remarkable relation may always be observed between the quantity of oxygen employed by the animal, and that of the carbonic acid it produces : under ordinary circumstances, the volume of the latter is but little below that of the former, and animals, as will be shown, all produce more or less heat.

There exists then the greatest analogy between the principal phenomena of respiration, and those of the combustion of charcoal : and this similarity of result has given rise to the idea, that the cause of the two might be the same.

And in truth one can hardly suppose the respiration of animals to be other, than the combustion by the oxygen of the air of a certain quantity of carbon, furnished by the bodies of these beings.   .

But where does this combustion take place ? Source of the carbonic acid. Is it the blood, which furnishes to the air the carbon thus consumed, and does this combustion take place on the surface of the respiratory organ? or, is the oxygen absorbed, and conveyed by the blood into the interior of all the organs, and the carbonic acid thus formed in all these parts, to be afterward expelled by the same way, which afforded a passage to the absorbed oxygen?

The majority of physiologists have adopted exclusively one or the other of these opinions ; but neither of these hypotheses is alone sufficient for the explanation of all the facts observed, and it would really appear, that the transformation of oxygen into carbonic acid takes place, both at the expense of the blood, at the moment of con-

11

tact of this liquid with the air, and in the substance of
the tissues, which compose our various organs : the following experiments may be adduced in proof.

If venous blood be enclosed in a flask filled with oxygen, and agitated, it is seen to change its color ; a part
of the oxygen disappears, and carbonic acid is próduced.
All the chemical phenomena of respiration, consequently
take place independently of life, and by the simple fact
of the contact of the blood with the oxygen. Now in
the bodies of respiring animals the blood is separated
from the air only by very thin membranes, which do not
at all oppose the contact. If phosphorus dissolved in oil
be injected into the veins of a dog, this substance, when
traversing the capillary vessels of the lungs, will combine
with the oxygen of the air, burn, and be driven out as a
thick white smoke. The blood must then evidently be
submitted in the respiratory organ to the contact of the
air, and furnish carbon to the oxygen of this fluid, as in
the experiment previously related ; and consequently it
must be acknowledged, that the direct combination of
the oxygen of the air with the carbon of the blood is
the source, at least of a part, of the carbonic acid produced.

But on the other hand, if an animal capable of resisting asphyxia for some time, a frog, for instance, be placed
in a vase containing no oxygen, and filled with azote, it
will continue to exhale carbonic acid, as if respiring air.
Now, in this case, it is impossible to attribute the formation of this gas to the direct combustion already spoken
of, for this combustion must cease as soon as the respired
air is deprived of its oxygen ; wherefore the carbonic
acid has simply been exhaled from the respiratory organ,

and formed elsewhere, from the oxygen already existing in the interior of the body of the animal.

The vapor, which escapes from the body at Source of the vapor expelled. the same time with the carbonic acid, also arises from the blood, and is simply exhaled from the surface of the respiratory organ. Some authors think, that this liquid is always formed during respiration, and that a part of the oxygen employed, by direct consumption of the hydrogen furnished by the blood, gives origin to the water; and thus they imagine that they can explain the cause of pulmonary transpiration, as well as the disappearance of a volume of oxygen superior to that of the carbonic acid formed. But experiment overthrows this hypothesis, for pulmonary transpiration continues, when the respired air no longer contains oxygen; and the quantity of vapor thus exhaled may be augmented at will, by the injection of water into the veins of a living animal.

All the volatile substances contained in the blood, are also expelled from the body by the exhalation from the respiratory organ. If camphor, or spirit of wine, be injected into the veins of a dog, they will soon escape with the vapor issuing from the lungs, and be recognised by their odor: the same result follows the injection of hydrogen gas in small quantities into the veins.

We have elsewhere seen, that the same organs also absorb with great rapidity the matters with which they are in contact; and this absorption is exercised upon the gasses, as well as upon the liquids; for an example take the following.

In one of the experiments made upon himself by the physiologist, Linning, he found, that his body had increased in weight eight ounces, without having made use of any aliment, and solely because he had respired an air charged with thick fog. Now, phenomena analogous to those thus accidentally manifested, take place normally in the ordinary performance of respiration.

Recapitulation. By a review of what has been said upon the nature of the respiratory function, it is found to consist in these four particulars, viz. :

1st. In the direct combustion of a certain quantity of the carbon of the blood by the oxygen of the air ;

2nd. In the absorption of oxygen, and exhalation of carbonic acid ;

3d. In the absorption and simultaneous exhalation of a small quantity of azote ;

4th. In the exhalation of water furnished by the blood, as are all the other expelled products.

Activity of the respiration. We have seen, that respiration is necessary for the support of life in all beings ; but the degree of activity of this function varies much in different animals.

Birds. Birds are of all animated beings the most active in their respiration ; in a given time they consume more air than any other class of animals, and therefore yield more readily to asphyxia.

Mammiferæ. The mammiferæ have also a very active respiration ; and a great number of experiments have been made, to estimate the quantity of oxygen that one of them, man, thus employs in a given time. This quantity varies with the individual, age, and various other circumstances ; but the average appears to be about seven hundred and fifty litres or cubic decimetres a day. Now oxygen forms only the twenty one hundredths (in volume) of the atmospheric air ; it follows then that man consumes, during this space of time, at least three thousand five hundred litres or cubic decimetres of the latter fluid.

Animals of the inferior classes have, in gene- *Inferior ani mals.* ral, a respiration much more limited, especially those living in the water. But yet when we reflect upon the enormous quantities of oxygen, that all these beings must consume daily, it is evident that the atmosphere would soon be deprived of it, and all animals perish asphyxiated, if nature did not employ certain means for the constant renewal of the quantity of this gas, diffused around the surface of the globe.

This she accomplishes by *the respiration of* *Influence of animals and* *plants;* and it is worthy of observation, that the *vegetables up- on the compo-* mean employed is a phenomenon of precisely *sition of the atmosphere.* the same order with that, whose effects it is intended to counterbalance.

Vegetables absorb the carbonic acid diffused in the atmosphere, and under the influence of the solar light they extract from it the carbon, and give out oxygen. Thus the vegetable kingdom supplies animals with the oxygen necessary for them, and the respiration of animals constantly furnishes vegetables with the carbonic acid necessary for their growth.

We thus see, that it is in a great measure upon the relation existing between animals and vegetables, that the nature of the atmosphere depends; and that in its turn the composition of the air must in some sort govern the relative proportion of these beings.[1]

_____

[1] From this it might be supposed, that in cities, where a great many men are collected, and where there are very few plants, the atmosphere must be less rich in oxygen than in the country, but it is an error. Chemical analysis demonstrates that the air has everywhere the same composition, and this uniformity must be attributed to the currents, by which the atmosphere is continually agitated.

Relation be-
tween the ac-
tivity of respi-
ration, and the
vivacity of
motion.
There always exists a remarkable relation between the quantity of air consumed by each animal in a given time, and the vivacity of its motions. Animals, whose motions are slow and rare, have, all other things being equal, a respiration much less extensive than those which move with rapidity, and remain but a short time at repose. A frog, or toad, for example, consumes far less air than certain butterflies, although its body be much larger than that of these insects ; but these reptiles move seldom and slowly, while the butterflies constantly execute the most lively motions.

Circumstances
influencing the
extent of the
respiration.
The activity of respiration varies also in the same animal according to the circumstances, in which it is placed ; and it may be established as a general proposition, that everything, which tends to diminish the energy of the vital movement, determines a diminution either in the absorption of the oxygen, or in the relative proportion of the carbonic acid exhaled ; while, on the other hand, everything, which augments the force of the animal, produces a corresponding change in the extent of the respiration.

Thus in the young this function is less extensive than in the same beings at the adult age.

During sleep the extent of respiration is also diminished. Fatigue, abstinence, the abuse of spirituous liquors produce the same effect. Moderate exercise and the taking of food exert on this function a contrary influence.

Finally, heat augments the extent of respiration and cold diminishes it.

It appears, that there exist also variations in the quantity of carbonic acid produced at different parts of the day, and, from some facts, it would seem that the pressure of the barometer exercises a very marked influence upon this phenomenon.

Hitherto we have been occupied merely with the phenomena of respiration, considered in themselves, and without regard to the organs, which are the seat of them. Let us now see what are the instruments of this important function, and how they are modified in different animals.

In those with the simplest organization, respi- Skin. ration is not assigned to any special apparatus, but is effected in all the parts, which are in contact with the element, in which these beings live, and from which they derive the oxygen necessary to their existence.

The general envelope of the body, or *the skin*, is likewise the seat of a respiration more or less active, in the greater part of animals of the more elevated classes, and especially in man ; but in all these beings a determined part of the tegumentary membrane is more especially destined to act upon the air, and is so modified in its structure, as to better fulfil this function.

In the animals, in whom respiration once be- Special organs. gins to localize itself, it has for its instrument a certain number of membranous appendages, which are raised upon the surface of the skin in some part of the body, and assume the form of tubercles, folds, or fringes.

In other animals, in whom respiration is more active, the portion of the general envelope of the body to which is assigned the performance of this act, in place of rising upon the surface, folds inward and constitutes sacs or canals, into which the air penetrates.

General char-
acters of the
respiratory or-
gans.   Whatever be the form, that the respiratory apparatus assumes, it is remarked, that the part thus modified to act upon the air, presents a soft, spongy, and fine texture ; that it receives a great quantity of blood ; and that it is arranged so as to offer, under a volume comparatively small, an extent of surface proportioned to the activity of the respiration. It may be established, as a general proposition, that this organ will be an instrument of so much the greater power, as its organization differs from that of the general envelope of the body, and that the respiration, which takes place by the skin, will be less active in proportion as that by these special organs is extended.

Differences
with regard to
the mode of
respiration.   The structure of the respiratory organs varies, according as they are to be in contact with the air in the state of gas, or to act upon water holding in solution a certain quantity of this fluid.

In all animals living under the water, and respiring by the intervention of this liquid, the special instruments of respiration are salient, and bear the name of *gills* ; while in animals with aërial respiration there are no gills but interior cavities answering the same purposes, and which are called the *lungs*, or *tracheæ*.

Gills.   The GILLS, in their simplest form, consist merely of a few tubercles with a texture a little softer than that of the rest of the skin, and which receive a slightly increased quantity of blood ; but they are very far from being the sole instruments of respiration, and the rest of the skin takes an active part in its performance.

Several marine worms possess this mode of organization ; but when these organs are to be the seat of a more active respiration, their structure is complicated, and they take the form of lamellæ very thin and numerous, or of simple, or ramified membranous filaments.

The former of these modes of structure is met with in most of those animals, which, with the crabs and lobsters, constitute the group to which the name of crustacea has been given, and in a great number of those inhabiting the interior of shells, and which constitute the class mollusca ; oysters, for example. The second modification of the gills is found in fishes, &c.

The interior cavities, which serve for the aërial respiration, sometimes take the form of tracheæ, sometimes of lungs.

The TRACHEÆ are vessels, which communicate   Tracheæ. with the exterior by openings called stigmata, and ramify in the interior of the various organs. They convey to them the air, and consequently respiration is effected in all parts of the body. This mode of structure is peculiar to insects and some of the arachnidæ.

The LUNGS are sacs more or less divided into   Lungs. cellules, which receive the air into their interior, and the walls of which are traversed by the vessels containing the blood to be submitted to the vivifying influence of the oxygen.

Lungs exist (but in a state of great simplicity,) in most spiders, in some of the mollusca, such as snails. Reptiles, birds, and the mammiferæ, are also provided with them.

In man, (as well as in all the mammiferæ,) Lungs of man. the lungs are lodged in the cavity called the *thorax*, which occupies the superior part of the trunk, and which is separated from the abdomen (or belly) by a transverse partition, formed by the *diaphragm*. These organs are, so to speak, suspended in this cavity, and are enveloped by a thin, close membrane, which also lines the thorax, and which is called the *pleura*.[1] They are in number two, one upon either side of the body, and they commu_nicate externally by means of a tube, the *trachea*, which ascends along the anterior part of the neck, and opens into the posterior fauces.

Fig. 16.[2]

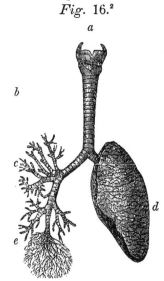

This duct is formed by a se-ries of small cartilaginous bands placed across, in the form of in-complete rings: their interior is lined by a mucous membrane of the same nature with that of the mouth, and continuous with it. Finally, at its inferior part the trachea divides into two branches, which take the name of *bronchi*, and which ramify into the inte-rior of each lung, as the roots of a tree in the soil.

The lungs, as we have already

---

[1] The disposition of the pleura is analogous to that of the other serous membranes, of which we have spoken (page 71). It forms a sac without opening, which is folded upon itself, and the external half of which adheres to the walls of the thorax, while the other half is fixed upon the surface of the corresponding lung: the internal face of the pleura is consequently every where in contact with itself, and as it is extremely smooth, and con-tinually lubricated by serosity, it slides very readily, and essentially favors the respiratory motions.

[2] This figure represents the trachea and the lungs; one of these organs has been left untouched (*d*), but on the other side the substance has been destroyed to expose the ramifications of the bronchi (*c*). *a*, Larynx and superior extremity of the trachea; — *b*, trachea; — *c*, divisions of the bron-chi; — *e*, bronchial twigs; — *d*, one of the lungs.

said, present in their interior a multitude of cellules, into each of which a small branch of the corresponding bronchus opens. The walls of these cavities are formed by a very fine, soft membrane, pierced by numerous capillary vessels, which receive the venous blood coming from the pulmonary artery, and expose it to the action of the air.

Within the same volume, the smaller the size of the cellules of which the lungs are formed, the greater is the surface, upon which respiration will act, and the more numerous are the points where the blood comes in contact with the air. There consequently exists a direct relation between the size of the pulmonary cellules, and the activity of the respiration; for example in frogs, in whom this function is exercised only in a slow and feeble manner, the lungs have the form of sacs, merely divided by a few partitions; while in the mammiferæ and birds, in whom respiration is most active, these organs are divided into cellules so small, that by the naked eye it is difficult to perceive them.

In man and the other mammiferæ, the bronchi all terminate in the pulmonary cellules, and these latter are always terminated in a cul-de-sac; wherefore the air, which enters the lungs of these animals, can advance no farther. But in birds, in whom respiration is yet more active, some of these canals traverse the lungs completely, and open into the cellular tissue which surrounds them, and which in the rest of the body fills the spaces between the different organs; the cavities contained in this tissue all communicate together, and the air, which arrives in them also penetrates into all parts of the body, even into the substance of the bones.

MECHANISM OF RESPIRATION.

From what has been said of the alterations, which the air undergoes from respiration, it is evident, that this fluid must be unceasingly renewed in the interior of the lungs ; which is effected by the movements of inspiration and expiration, that we momentarily execute.

The mechanism, by which the air is drawn into the lungs, or expelled from them, is very simple, and resembles in all points the play of a bellows, except that in the former the fluid penetrates into and escapes from the organ by the same outlet. The walls of the thorax being movable, its cavity may be alternately enlarged and contracted, and the lungs will follow all their movements ; thus in inspiration, the column of air pressed by the whole weight of the atmosphere, is driven into the chest through the mouth, nasal fossæ, and trachea, and fills the pulmonary cellules, in the same way that water ascends in the body of the pump when the piston is raised. In expiration, the air contained in the lungs is, on the contrary, compressed and driven out by the same way, which served for its entry.

To understand how the thorax of man dilates and contracts, it is necessary to examine its structure.

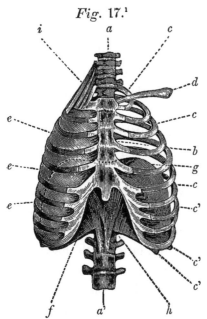

Fig. 17.¹

This cavity ^{Structure of the thorax.} has the form of a cone with the apex above and base below, and its walls are chiefly formed by a kind of bony cage, resulting from the union of the *ribs* with a portion of the *vertebral column* (or spine,) behind, and with the *os sternum* in front.

The spaces between the ribs are filled by muscles, which extend from one of these bones to the other ; muscles also pass from the first rib to the cervical portion of the vertebral column ; lastly, the inferior wall of the chest is formed by the *diaphragm*, which is attached to the inferior border of the bony case just mentioned.

The dilatation of the thorax may take place ^{Dilatation of the thorax.} in two ways, by the contraction of the diaphragm, or elevation of the ribs.

¹ Thorax of a man. The muscles of the left side are removed, those on the right entire.

a, Cervical portion of the vertebral column ; — *a'*, lumbar portion of the column ; the dorsal portion, which enters into the formation of the thorax, is concealed by the sternum, &c. ; — *b*, sternum ; — *c, c*, ribs ; — *c'*, false ribs ; — *d*, clavicle ; — *e*, intercostal muscles ; — *f*, lower false rib, concealed by the insertion of the diaphragm ; — *g*, arch formed in the interior of the thorax by the diaphragm ; on the right side the continuation of this arch is indicated by the dotted line ; — *h*, pillars of the diaphragm inserted into the lumbar vertebræ ; — *i*, levator muscles of the ribs.

Contraction of the diaphragm. The diaphragm in the state of repose forms an elevated arch, which ascends into the interior of the chest (*g*) ; and it is easy to perceive, that the contraction of this muscle must diminish the curvature of the arch, and thus proportionately enlarge the cavity of the thorax.

The action of the ribs is a little more complicated ; these bones, (*c* and *c'*) to the number of twelve on a side, describe each a curve, the convexity of which is turned outwards and a little downwards ; their anterior extremity, which is united to the sternum (*b*) by means of intervening cartilages, is much less elevated than the posterior, and the articulation of the latter with the vertebral column, permits them to ascend and descend. The former of these motions is determined by the contraction of the muscles at the base of the neck, (*i.*) When the ribs ascend, their tendency is to form a horizontal line ; for at the same time that their anterior extremity ascends, carrying with it the sternum, they turn a little upon their axis, so that their curve is no longer directed downward, but outward ; wherefore the lateral and anterior walls of the thorax are removed to a greater distance from the vertebral column, and the capacity of the chest increased.

Motion of expiration. In the movement of expiration the diaphragm is relaxed, and the lungs, by reason of the elasticity of their tissue, contract, and draw with them this muscular partition, till it ascends like an arch. When the muscles, which have produced the elevation of the ribs and sternum, cease to contract, the weight of these bones, and the traction made by the elasticity of the lungs causes their depression ; but there are also other forces, which may contribute to the diminution of the capacity of the thorax, and the expulsion of the air from the lungs : such are the contractions of the muscles, which form the walls of the abdomen, and which are fixed to the inferior part of the chest.

Several degrees may be remarked in the ex- Capacity of the lungs. tent of these movements, and in ordinary respiration, the quantity of air aspired by the thorax, or driven from the lungs, scarcely exceeds the seventh part of what these organs can contain. The quantity of air, contained in the lungs under ordinary circumstances, has been estimated at nearly 4580 cubic centimetres, and that which enters the chest, or issues from it, at each inspiration, or expiration, at 655 cubic centimetres.

The number of the respirations varies with Frequency of the respirations. the individual and the age ; in infancy they are more frequent than in the adult, and in the latter there are about twenty inspirations to the minute.

Thus we see, that in the ordinary state there must enter the lungs of a man about 13,100 cubic centimetres of air in a minute, making about 786 litres to the hour, and the daily consumption nearly 19,000 litres of this fluid.

Sighing, *gaping*, *laughing*, and *sobbing*, are Modifications of the respiratory motions. but modifications of the ordinary movements of respiration. *Sighing* is a long and deep inspiration, in which a great quantity of air gradually enters the lungs ; thus this phenomenon does not depend solely upon the moral affections, which are its most frequent cause, but the need of sighing is felt whenever the respiratory function is not effected with sufficient rapidity.

Gaping is an inspiration yet deeper, accompanied by

an almost involuntary and spasmodic contraction of the muscles of the jaw and velum palati.

Laughing consists in a succession of slight interrupted expiratory motions, varying in frequency, and chiefly dependent upon the almost convulsive contractions of the diaphragm. The mechanism of *sobbing* differs but little from that of laughing, although the former expresses very different affections of the mind.

THE INFLUENCE OF RESPIRATION UPON THE OTHER FUNCTIONS.

In concluding our remarks upon respiration, we will add a few words upon the influence of its various movements on the other functions, the history of which has already been given.

Upon the circulation. It is evident, that the dilatation of the thorax must produce upon the blood, contained in the great vessels emptying into this cavity, the same effect as upon the air contained in the trachea. By the movements of inspiration, that portion of the venæ cavæ, comprised within the thoracic cavity, is swollen by the access of blood thus aspired ; and from the same cause the veins, which enter this cavity, but are situated externally, and consequently under the influence of atmospheric pressure, are more or less completely emptied.

This kind of suction contributes to aid the course of the blood in the venous system, and is felt even in the arteries, with which the former vessels are continuous through the intervention of the capillaries.

The motions of expiration, on the contrary, momentarily suspend the course of the blood in the great veins, and accelerate it in the arteries, which originate from the heart, and are thus compressed.

To these two phenomena must be attributed the swelling of the veins, (especially those of the head and neck), during expiration. In the interior of the cranium this swelling is so marked, that at each respiratory movement the vessels, situated at the base of the brain, elevate this viscus, and produce a kind of pulsation.

The dilatation of the chest appears also to Upon the absorption. exercise a remarkable influence upon the absorption ; in fact it acts like a pump upon all which surrounds the thorax, and must tend to force from without inwards, all the fluids which communicate with its interior ; but this action is felt only in the immediate neighborhood of the chest.

Finally, the abundant exhalation, which al- Upon the pulmonary exhalation. ways takes place upon the surface of the pulmonary cellules, is determined in a great measure by the degree of suction, which accompanies each movement of inspiration, and which acts upon the liquids, by which the walls of these cellules are moistened, as it acts upon the blood of the venæ cavæ, and primarily upon the air of the trachea. We have already seen, that in the normal state, all the volatile substances met with in the blood are thus exhaled ; but if the thorax of a living animal be opened, and artificial respiration performed, so that there is no suction upon the surface of the pulmonary cellules, this exhalation is almost entirely arrested ; and then camphor injected into the veins does not escape more rapidly in this way, than from the surface of every other membrane, whose tissue is as vascular and permeable to liquids.

13

ANIMAL HEAT.

There exists another phenomenon, the history of which
is closely united to that of the respiration, and the importance of which is so great that we must necessarily
devote some time to it, viz. : *the faculty of producing heat.*

Difference
in the temper-
ature of ani-
mals.
This faculty appears to be common to all animals; but the majority develop so little caloric,
that it cannot be appreciated by our ordinary thermome-
ters ; while in the remainder the production of heat is so
great, that we are not required to make use of scientific
instruments to prove its existence.   To compare this difference, we have only to place a hare and a fish, having
about the same volume, in two calorimeters, and to surround them with ice at the temperature of 0° ; the
quantity of this body melted in a given time will be proportional to the quantity of heat developed by these two
animals.   Now, in the instrument containing the fish,
the quantity of ice melted in the space of three hours,
for example, will not be appreciable ; while in that containing the hare, we shall find, after the same lapse of
time, more than a pound of liquid water; and to melt
this quantity of ice would require as much heat, as to
warm from melting ice to boiling water about three
fourths of this weight of water ; this heat could only have
been furnished by the animal under experiment.

This enormous difference in the production of heat
occasions corresponding differences in the temperature of
the different animals.   A thermometer placed in the
body of a dog or bird, for example, will always ascend to
36 or 40 degrees, (centigrade,) while in the body of a

frog or fish, it will indicate a temperature nearly equal to the atmosphere at the time of the experiment.

These are called *animals with cold blood*, which do not produce a sufficient degree of heat, to have a temperature of their own and independent of the atmospheric variations; those are *animals with warm blood*, which preserve a temperature nearly constant through the ordinary variations of heat and cold, to which they are exposed. Birds and mammiferæ are the only beings included in the latter class; all others are animals with cold blood.

The temperature of man and most of the other mammiferæ scarcely varies from 36 to 40 degrees; that of birds ascends to about 42 degrees, centigrade. *Temperature of the mammiferæ and birds.*

But the faculty of producing heat varies in the different animals of these two classes, and also in the same individual, according to age and circumstances. Thus most mammiferæ and birds produce a sufficient degree of heat, to preserve the same temperature in summer and winter, and to resist the ordinary effects of cold, though it be very severe. But there are others, which produce only heat enough to raise their temperature 12 or 15 degrees above that of the atmosphere; wherefore, during summer, their temperature is nearly the same with the warm-blooded animals, but in the cold season it is much diminished; and whenever this cold reaches a certain limit, the vital movement is rendered slower, and the animal experiencing it falls into a state of torpor, or lethargic sleep, which lasts till the temperature again ascends. The beings, which present this singular phenomenon, are called *hibernating animals*, and in this re- *Hibernating animals.*

spect they are in some sort intermediate between the
warm-blooded animals non-hibernating, and the animals
with cold blood.

Influence of age upon the production of heat. In the early periods of life, all warm-blooded
animals are more or less nearly allied to the
cold ; these, as well as the latter, do not produce in general sufficient heat to preserve their temperature, even
when exposed to very slight degrees of cold. But the
decrease of temperature, which is without inconvenience
for the cold-blooded animals, acts upon the others in a
very different manner, for always, if carried beyond a
certain degree, or lasting a determined period, death is
the result. With regard to the faculty of producing heat,
the young of warm-blooded animals, which are born with
their eyes open, and which immediately after birth can
run and seek their nourishment, differ far less from the
adults, than the mammiferæ, which are born with their
eyes closed, or birds, which on issuing from the egg, are
not yet covered with feathers. If, for example, new
born cats and dogs are taken away from the parent, and
exposed to the air, even in summer, they are chilled to
the very point of death.

Infants produce also much less heat in the first days
following birth, than at a more advanced period of their
life ; their temperature then decreases very readily, and
the influence of cold is very injurious to them ; therefore
a greater number die in winter than during the rest of
the year.

Everything, which acts as an excitant and which augments the energy of the vital movement, tends also to
augment the faculty of producing heat, and everything,
which weakens the animal economy, exercises a debili-
tating influence upon this function.

Thus the action of a moderate cold tends to _{the temperature.} *Influence of the temperature.* augment the faculty of producing heat, and in couse- quence during winter we can better resist the causes of chill than in summer.

The influence of heat, when not prolonged for too long a period, is excitant and increases the faculty of producing caloric ; but, if long continued, it weakens the body and diminishes the energy of this faculty. For this reason, persons, who have resided for some time in tropi- cal regions, are so sensible to the cold of our winters.

Finally, exercise ever augments the produc- *Influence of exercise, &c.* tion of heat, and the acceleration of the respiratory move- ments is followed by the same effect. During sleep, this faculty appears to be, on the contrary, less powerful than when awake ; thus when men, exposed to a very low temperature, have the imprudence to fall asleep, they yield to its effects more rapidly than if awake and in motion. The disastrous retreat from Russia furnishes numerous examples of the bad effects of sleep upon the soldiers, weakened by fatigue and privation of all kinds, and exposed to a most intense cold.

The cause of the production of heat in the *Cause of the production of heat.* body of animals appears to be the action of the arterial blood upon thc tissues, under the influence of the nervous system. There exists an evident relation be- tween the faculty of producing heat, the intensity of the nervous action, the richness of the blood, and the more or less rapid transformation of the venous to arterial blood.

Experiment has proved, that every thing, *Influence of the nervous system.* which tends considerably to-weaken the action of the nervous system, tends also to diminish the produc- tion of heat. Thus if the brain or spinal marrow of a

dog be destroyed, and the respiration be imitated by artificial means, the life of the animal is indeed sustained,
but the production of heat ceases, and the body becomes
cold as rapidly as a dead body would do, if placed in
similar circumstances. If the action of the brain be
paralyzed by certain energetic poisons, such as opium,
the same effect is produced, and these experiments varied
in different ways have firmly established the fact, that
one of the conditions necessary to the development of
animal heat is, the influence exercised by the nervous
system upon the rest of the body.

Influence of
the blood.   On the other hand, the action of the blood
upon the organs appears to be equally indispensable to
the manifestation of this phenomenon ; for, the suspension of its circulation in any part of the body is followed
by coldness of the part ; and, moreover, a remarkable
relation exists between the faculty of producing heat in
different animals, and the richness of their blood.   Birds,
which have the highest temperature of all animals, are
also those whose blood is the most loaded with solid particles (in general fourteen or fifteen parts to the hundred);
the mammiferæ, whose temperature is not quite so elevated, have more aqueous blood, in general the weight of
the globules constituting only the nine or twelve hundredths of the total weight of this liquid ; lastly, in the
cold-blooded animals, such as frogs and fishes, we find
barely six parts of globules and ninety-four of serum.

Influence of
the respiration.   But the action of the nervous system, and of
a blood more or less rich in globules, are not the only
circumstances, which influence the production of animal
heat ; in order that the nutritive liquid may exercise upon
the economy the necessary action, it must possess all the

properties, which characterize the arterial blood ; and as it acquires these only from respiration, the development of caloric must also depend upon this latter function. All the causes, which render the transformation from venous to arterial blood less complete, or less rapid, tend also to diminish the production of the heat, and there always exists an intimate relation between it and the activity of the respiration.

The formation of cabonic acid, which is one of the most remarkable phenomena of the respiration of animals, may also explain to us the cause of the production of the greater part of the heat developed by these beings. If the oxygen absorbed during respiration is employed to form this gas by its union with the carbon, arising from the blood or from the living tissues, as we have every reason to suppose, this combination must be accompanied by a disengagement of heat as from the combustion of charcoal in the air.

Numerous experiments, made with an extreme precision, demonstrate that the heat, which would be produced by the combustion of the carbon contained in the carbonic acid gas exhaled by warm-blooded animals, is equal to more than half the quantity of caloric disengaged by these beings. And if we admit, that the absorbed oxygen, without being replaced by the carbonic acid, combines in the interior of the body with the hydrogen to form water, we see, that the heat produced by this combustion taken with that of the carbon already mentioned, would be equivalent to nine tenths of that developed by the animal. The motion of the blood, and the friction of the different parts of the body probably produce the remainder.

As a last analysis we see then, that the respiration is the principal cause of the production of animal heat; and that the kind of combustion, occasioned by the action of the oxygen upon the blood and living organs, is effected solely through the influence of the nervous system.

Temperature of the different parts of the body. This important function is not, however, exercised with the same energy in all parts of the body; those, in which the blood circulates, with greater abundance and rapidity (and in which, cousequently, life is the most active), are also those, in which the most heat is disengaged; and therefore the organs more distant from the heart, must, cœteris paribus, produce less heat, and consequently be more readily chilled. This is the true state of the case; the temperature of our limbs being less elevated than that of the trunk, if we are exposed to the action of an intense cold, these parts are the first to freeze.

Power of resisting heat. The faculty of producing heat explains to us, why animals with warm blood have a temperature, which can sustain itself above that of the surrounding atmosphere. But how happens it, that these beings can preserve the same temperature, when they are placed in air hotter than their body? A man, for example, can remain during a certain time in a dry hot-house, where the air is heated to a degree approximating boiling water, without any perceptible increase of the temperature of his body except two or three degrees.

The power of thus resisting heat depends upon the evaporation of water, constantly going on at the surface of the skin, or in the apparatus of respiration, and which constitutes the *cutaneous and pulmonary transpiration ;* for the transformation of water into vapor, withdraws caloric from every thing surrounding it, and the body is chilled as fast as warmed by the external heat.   For the same reason water placed in the porous vases, called *alcarazas*, cools so promptly even in mid-summer.[1]   Now the quantity of water thus evaporated, increases with the temperature of the air, and the cause for the cooling becomes more powerful, the greater the heat of the atmosphere.

————————

DIGESTION.

We have already seen, that all living beings are obliged continually to seek in the exterior world for nutritious substances, and to assimilate to their organs new materials.   We have also seen how this absorption is effected, and the study of respiration affords us examples of these substances thus penetrating into the nutritive liquid, and being carried by it into the interior of the organs, without having undergone any previous modification.

In vegetables, all nutritive substances penetrate directly into the organs.   But, in animals, the greater part of the materials necessary for the support of life, are not

———

[1] These vases allow the water they contain to filter through, and thus have a constantly moist surface, from which a rapid evaporation takes place, which cools the liquid contained in their interior.   From the same cause, do we experience so lively a sensation of cold, when ether is poured upon the skin, and the part thus moistened blown upon.

14

absorbed until they have undergone a certain preparation, by means of which their properties are changed, and their composition modified ; or, in other words, till after they have been DIGESTED.

Aliments.   The name of *aliments* may be given to all substances, which, introduced into the body of a living being, serve for its increase, or to repair the losses it constantly undergoes ; but in general, the sense of this word is more restricted, and is only applied to the materials which are not absorbed, and may not serve for nutrition, till after they have been digested.   For the sake of perspicuity, we shall only employ it in the latter acceptation.

Phenomena depending upon the lack of aliments.   Aliments are not less necessary to life, than the air we breathe, or the water which our body continually absorbs, either in the liquid state as drink, or in the form of vapor.   When animals are deprived of them, their body diminishes in volume, their powers are weakened, and death arrives after sufferings more or less prolonged.

Hunger.   The want of aliments first makes itself known by a peculiar sensation, which has its seat in the stomach, hunger.   It is increased by exercise, by the stimulating influence of a moderate cold, and by the action of certain bitter substances, as the cashew nut, upon the stomach.   On the contrary, everything, which retards the vital movement, immobility, sleep, &c., also tends to render this want less imperious.   Hibernating animals take no food during the season of their lethargy, and cold-blooded animals, such as fishes and frogs, can support a very long abstinence, when the exercise of their functions is retarded by the influence of a very low température. But the animals, whose nutritive movement is very rapid, such as man and most mammiferæ, soon perish for the want of food. The herbivorous, whose blood is less rich in globules than the carnivorous, yield sooner than the latter; and young animals, whose nutrition is much more active than in the adult (since the volume of their body is constantly increasing, in the place of remaining stationary), also die of hunger sooner than the latter. Dante, in the famous episode of Count Ugolino, has but depicted in lively colors the truth, as it would occur, if a man, already having reached the period of his growth, and children of a tender age, should at the same time be deprived of all nutrition.

Prolonged abstinence occasions very remarka- Death from inanition. ble phenomena, which may be ranked under three heads. In the first period, hunger makes itself frequently felt, and occasions a greater or less degree of weakness, with considerable alteration of the features. In the second period, the intellectual faculties are troubled; in man, as well as in animals, inquietude, or even fury arises, and sometimes mental alienation is manifested by visions. Lastly, in the third period, this exaltation gives place to a state of depression or complete stupidity, and it is to be observed, that often when abstinence has been prolonged beyond a certain period, the use of aliments can no longer save the life of the individual. In this latter case death almost always ensues, whether the animal continue to fast, or retake its ordinary regimen.

The aliments are all furnished by the organic Nature and properties of kingdom, and life is maintained in man and all the aliments. other animals from substances, which have themselves made part of a living being.

All alimentary substances do not possess the nutritive property in the same degree, and very curious experiments have established the fact, that in most animals at least, the union of a certain number of different substances is indispensable to the wants of life. Thus hares, nourished upon a single substance, such as cheese, cabbages, oats, or carrots, die in the space of about fifteen days, with all the appearance of inanition, while if nourished with these same substances, given together, or successively at short intervals, they live and do well.

The diversity and multiplicity of the aliments, is then an important rule of hygiene; and in this, the precepts of science agree perfectly with our instinct, and with the variation, brought by the seasons in the alimentary substances offered us by nature.

Experiment has also proved, that substances, such as sugar, gum, oil, and fat, into the composition of which no azote enters, cannot suffice for the nutrition of animals, in whatsoever order they may be given. The use of a certain quantity of azoted aliments, such as muscular flesh, the gluten found in the grain, albumen, &c., appears to be indispensable to the support of the life of all animals.

When we compare the nutritious qualities of the different alimentary substances, we must also take into consideration the quantity of water they contain; by deducting this from the weight of the mass employed, we arrive at the knowledge of the really nutritious matter. Thus our ordinary bread contains, to the 1000 parts, 250 of water; beef about 700; potatoes 750; turnips and cabbages 950.

But the different substances, which may serve as

aliments to animals, vary with the nature of these beings ; and these differences, as we shall see, are always in relation to other differences in the organization. From an investigation of the digestive apparatus, we can understand why one animal is nourished upon vegetables, and another upon flesh. But one thing, for which we cannot account, but which is notwithstanding very true, is the faculty possessed by certain animals of nutrition upon substances, which, to others, are violent poisons. Thus, goats and sheep may eat with impunity hemlock, while a very small quantity of this plant is sufficient to destroy man and many other animals.

Digestion, or the process by which animals *Modification of the aliments by digestion.* modify the aliments so as to render them proper for absorption and nutrition, consists essentially in the action of certain humors upon these matters, in consequence of which they undergo various alterations, and are separated into two parts ; one destined to penetrate the interior of the body, and supply the wants of the animal, called *chyle*, the other improper for this purpose, and ultimately expelled under the form of *fæces*.

From the nature of this process, it is evident, *Seat of the digestive process.* that digestion must always take place in an interior cavity of the body, which may serve as a reservoir for these humors, and for the aliments they are to act upon. All animals are provided with a *digestive cavity*, and the existence of this organ is one of the characters distinguishing them from vegetables, in which the alimentary substances are absorbed without any previous preparation.

In some animals with very simple structure, this sac is but a fold of the skin which penetrates deeply into the

body, and terminates in a cul-de-sac. This is the case
with the hydræ or fresh water polypi, of which we have
already spoken : thus one of these animals can be turned
like the finger of a glove, without changing its manner of
living. The surface, which was exterior, then becomes
interior and forms the cavity, in which the aliments are
digested, while the surface which originally lined this
cavity, becomes external, and no longer acts upon these
substances.

Digestive apparatus. The digestive cavity of man and most an-
imals has the form of
a long canal, extending
from one extremity of the
trunk to the other, with
alternate dilatations and
contractions, so as to constitute several kinds of
chambers or pockets, uni-
ted by ducts more or less
narrow. This tube is
formed by a *membrane*,
called *mucous*, analogous
in structure to the skin,
with which it is continu-
ous; differing however in
its greater degree of soft-

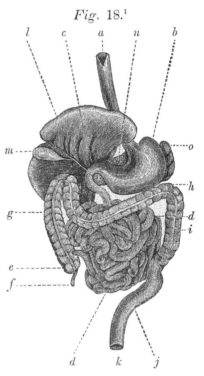

Fig. 18.[1]

---

[1] Digestive canal and its appendages.

a, Æsophagus, — b, stomach, — c, pylorus continuing with the duodenum, or first portion of the small intestine, — d, d, small intestine, — e,
cæcum, or first portion of the large intestine, in which the small intestine
terminates, — f, vermiform appendix of the cæcum, — g, ascending colon,

ness, the greater number of its capillary vessels and secretory follicles, and in the almost complete absence of epidermis. Surrounding this membrane is a fleshy envelope,
formed of *muscular fibres*, somewhat abundant, and by their
contractions driving the alimentary substances from the
mouth to the anus, or arresting them in their course, and
impeding their progress for a certain time in one, or another part of the digestive apparatus. Lastly, in a great part
of its extent, this tube is also enveloped by a serous
membrane, thin, and transparent, called the *peritoneum*,
which serves both to fix it, and to facilitate its movements.

The digestive apparatus is composed of this alimentary tube, of the organs destined to divide the aliments,
of the various glands serving to form the humors necessary for digestion ; and of the vessels charged with the
duty of conveying the nutritive materials thus elaborated,
from the digestive cavity into the interior of the apparatus of circulation.

The alimentary tube takes, in various parts, different
names. Its anterior part, enlarged, and filling the purpose of a sort of vestibule, is called *the mouth*, the cavity
continuous with it is called the *pharynx ;* the third part
of the canal constitutes the *œsophagus ;* the fourth the
stomach ; the fifth the *small*, and the sixth the *large intestines*, which terminate at the *anus.*

In man, and the animals nearly allied to him, the
organs which effect the mechanical division of the aliments, are situated in the mouth, and are called *the teeth.*

— *h*, transverse colon, — *i*, descending colon, —*j*, rectum, — *k*, extremity
of the rectum, — *l*, liver, — *m*, gall bladder, — *n*, pancreas, a great portion
of this gland is concealed behind the stomach, — *o*, spleen.

But in certain animals this process is confided to other parts, to the stomach, for example, as in birds.

The principal glands of the digestive apparatus are ; the *salivary glands*, the *gastric follicles*, the *liver*, and *pancreas*.

Lastly, the vessels which serve for the absorption of the products of digestion, are in man, as well as all other mammiferæ, birds, reptiles, and fishes, special canals called *chyliferous*, or *lacteal vessels*.

All these organs, with the exception of the mouth, salivary glands, pharynx, and æsophagus, are lodged in a large cavity, which occupies the inferior two thirds of the trunk, and is called the *abdomen*, or belly. It is separated from the thorax by the diaphragm, and terminated inferiorly by a basin, formed of a large bony circle, the centre of which is occupied by a kind of fleshy wall. Behind, it is bounded by the spine, and in front, as upon the sides, its walls are formed by large muscles, which extend from the thorax to the basin, (*pelvis*,) of which we have just spoken. The internal surface of this cavity is lined by the peritoneum, and this membrane moreover forms various turns, between the folds of which are contained the stomach, intestines, liver, pancreas, and spleen. These folds, called *mesenteries*, all originate from the posterior part of the abdomen, and some among them are prolonged much beyond the organ they are to cover, and thus form veils or aprons called *epiploons*.

Prehension of the aliments. The introduction of the aliments into the digestive canal is effected in various ways ; and its mechanism is varied according as the substances are liquid, or solid ; but it is always performed either by the movements of the mouth, or by means of the superior extremities.

With anatomists, the *mouth* does not consist _{Mouth.}
in the opening which separates the two lips, but in the
oval cavity, formed above by the upper jaw and palate,
below by the tongue and lower jaw, laterally by the
cheeks, behind by the velum palati, and in front by the
lips. Its external communication may be enlarged or
closed at will, either by the movement of the lips, or separation or approximation of the jaws. It is then easy
to understand, how it may serve for the prehension of
aliments. These organs act as pincers, and seize the
bodies which are to be introduced into the mouth. In
most animals, these organs are situated anteriorly, to
seize the aliments; but in man, and some other animals,
this function is discharged by other members. The hand
places the aliments in the mouth, and the jaws approximate only to retain them there.

Liquids are taken in two ways; sometimes the liquid
is turned into the mouth, and falls by its own weight;
at other times, it is sucked in, either by the dilatation of
the thorax, which thus determines the entry of air into
the lungs, or by the movements of the tongue, which by
retreating backwards, acts in the manner of a piston;
the latter constitutes the phenomenon of sucking.

Fluids do not ordinarily remain in the mouth, _{Stoppage of}
but descend at once into the stomach; while _{in the mouth.}
the solid aliments remain in it a certain time, and are
submitted to *mastication* and *insalivation*.

Mastication, or the mechanical division of the _{Mastication.}
aliments is effected by the teeth.

These organs are bodies of an extreme dura- _{Teeth.}
bility, implanted in the border of either jaw, so as to act
upon each other. They greatly resemble bone, but dif-

15

fer in one important respect ; for bones are living parts, and constantly nourished, as shown by the experiments upon their coloration, while the teeth do not live ; they are not the seat of a nutritious movement, and the materials of which they are composed, are not renewed.   In this they resemble the hair, nails, and all the products secreted by the glands, such as the saliva, bile, and urine. Only in the place of being always liquid, as the latter, they soon become solidified, and acquire an extreme durability.

Mode of formation.   The teeth are formed by secretory organs contained in the interior of the jaws, (*d*, fig. 19). These organs are small membranous sacs, (*capsules, or matrices of the tooth*,) at the bottom of which is a small pulpy nucleus, called the *germ*, and in which ramify many nervous filaments, and a great number of blood-vessels, (fig. 20).   The bulb, or germ, (*b*,) permits a gelatinous humor to transude, which fills the capsule, (*a*,) and there is soon deposited upon its superior surface some grains of a stony substance, (*d, d*,) which enlarge by the exudation of a new quantity of matter, and unite so as to envelop the pulpy nucleus, from which they originate.   The solid envelope, resulting from this species of crystalization, is moulded exactly upon the germ, and as this is to con-

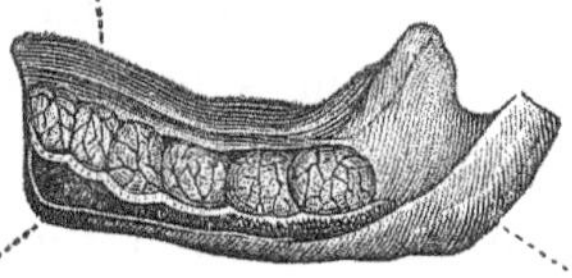

Fig. 19.[1]

---

[1] This figure represents the lower jaw of a very young infant; the greater part of the exterior surface of the bone has been removed to expose the capsules of the teeth in its interior ; *a*, gum ; *b*, inferior border of the jaw ; *c*, angle of the jaw ; *d*, capsules of the teeth ; *e*, coronoid process ; *f*, condyle of the jaw.

stitute the tooth, the form of these bodies must depend upon that of the germ itself. In proportion as this organ allows a new quantity of stony material to exude, the latter is gathered to that previously formed, and constitutes a new layer, situated upon the preceding.

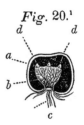

Fig. 20.[1]

The tooth thus enlarges by the addition of successive and concentric layers, and the germ is at last found contained in a canal, occupying the middle of this body, and which diminishes as new materials are interposed between this organ and the substance of the tooth. When the germ adheres to the bottom of the capsule only by a single point, the tooth can terminate only by one prong, or root; but if this organ adhere by several points, the stony matter secreted by it, penetrates between the peduncles, envelops the part beneath the germ, and by its prolongation forms as many prongs, or roots, as there are points of adherence.

Thus is the body of the tooth formed, and developed; but while the stony matter is thus being deposited by layers on its interior, the surface becomes encrusted by a still harder substance, which is formed by the capsule, and is called the *enamel*, while the central part secreted by the germ is called the *ivory*. Upon the superior part of the sac, enveloping the germ, a multitude of very minute vesicles may be observed, which are arranged with much order, and which secrete a peculiar liquor, which expands by minute drops upon the tooth,

[1] Section of the capsule of a tooth magnified to show the disposition of the germ, and the manner, in which the stony matter is deposited upon its surface; *a*, capsule; *b*, bulb or germ; *c*, blood-vessels and nerves, which penetrate to the bulb; *d, d,* first rudiments of the tooth.

thickens and forms the kind of varnish already mentioned.

In man and the carnivorous animals, the teeth are formed solely of these two substances, the ivory and the enamel; but in the herbivorous mammiferæ, some of these bodies present a third substance, which covers the enamel, and is therefore called the *cortical*; it is secreted by the capsule, and much resembles the ivory.

Chemical composition of the teeth. The ivory of the teeth is composed of gelatine, mixed with the phosphate of lime, (in the proportion of about sixty parts to the hundred in the adult man,) and containing also a small quantity of the carbonate of lime (ten to the hundred parts of ivory). The enamel contains only twenty to the hundred of animal matter, and eight of the carbonate of lime. According to some chemists there is also a florate of lime; but the existence of this material does not appear to be constant, and in any case it is only met with in small quantities. But the enamel is particularly distinguished from the ivory, by its compact and fibrous tissue, its color, and its hardness, which is so great that sparks may be elicited by collision with steel, like the flint.

Development of the teeth. As the tooth increases by the addition of new layers, either of ivory, or enamel, it approaches the edge of the jaw, then traverses it, issues from the gum with which this border is furnished, and appears externally; but the inferior part of the tooth which is of later formation, remains in the jaw and serves to fix it there. The name of *alveolæ* is given to the bony cavities in which the teeth are implanted, and that of *roots* to the parts enclosed: that part which appears above the gum is called the *crown*, and the *neck* is the connexion of the crown and root.

The roots differ also from the crown of the teeth by the absence of enamel, with which the latter is, on the contrary, covered ; and the cause of this difference evidently resides in the position of the part of the capsule, which secretes the stony varnish : it is in relation with the superior part of the teeth, but does not descend to the peduncle of the bulb, where the roots are formed.

The teeth present different forms, and their ^{Form of the teeth.} uses vary with the nature of these differences ; some terminate by a thin sharp edge, serving to cut the substances introduced between the jaws, and have received the name of *incisors*, (fig. 21, *a*, *b*,). Others are conical, and in most animals advanced beyond the neighboring teeth ; they do not serve to cut the aliments as the incisors, but to lay hold and tear them ; they are called the *canine teeth*, (*c*,). Lastly, others terminate by a broad, unequal surface, and present the most favorable conditions for crushing and grinding the aliments ; these are the *molar teeth*, or *grinders*, (*d, e, f, g, h,*).

In the study of animals, the disposition of the teeth is found to vary, according as these beings are to be nourished upon animal or vegetable substances, soft flesh, or little animals hid under a coriaceous or horny coating, as insects, tender herbs, or firm wood ; and to such a degree that by the mere inspection of these organs, one may obtain with much certainty a knowledge of the regimen, manners, and even general structure, of most mammiferæ.

The mouth of man is furnished with the three kinds of teeth already mentioned, and the manner in which they are planted in the jaw varies as much as the form of their crown. The incisors, (*a*, *b*,) whose action tends to bury them in their alveolæ, rather than tear them out, have but a single root quite short. The canine teeth, (*c*,) extend into the jaws more deeply than the incisors, and the molars, (*d, e, f, g, h,*) which must support the greatest efforts, present two or three diverging roots, to augment the solidity of their insertion.

Fig. 21.[1]

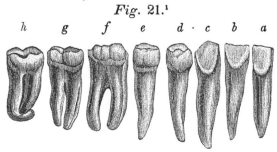

h   *g*   *f*   *e*   *d* · *c*   *b*   *a*

First denti-tion. At the period of birth, the development of the teeth is but little advanced; it is very seldom, that any of these bodies have yet pierced the gum, and their evolution does not commence till the age of six months, or a year. The teeth, which then form, are destined to fall out at the end of a few years, and to give place to others. They are called *milk-teeth,* or of the *first dentition,* and are twenty in number, viz.: in each jaw, four incisors, which occupy the front of the mouth, two canines, situated one on each side next the incisors, and four molars, placed at the bottom of the mouth, two on each side.

Second den-tition. The teeth of the *second dentition* are more numerous than those of the first; the complete number of these bodies is thirty-two, viz. : to each jaw, four incisors, two canine, and ten molars, the two first of which on each side have but two roots, and are called *small*

---

[1] Teeth of an adult; *a,* first incisor ; *b,* second incisor ; *c,* canine; *d* and *e,* small molars ; *f, g, h,* large molars.

molars, (fig. 21, *d, e,*) while the three next, situated at the bottom of the mouth, are provided with three roots, and called *large molars,* (*f, g, h,*).

In extreme old age, these teeth fall out as the milk-teeth of infancy, but are not replaced, and their alveolæ are obliterated.

The teeth, the development and structure of _{Muscles of mastication.} which we have been studying, are the passive instruments of mastication : they are moved with the jaws in which they are planted. The upper jaw does not move upon the rest of the head, but the lower, whose form resembles a horse shoe, articulates with the cranium only by the extremity of its two branches, and may be separated from, or approximated to the upper jaw. A great many muscles are fixed to this bone, and impress upon it their motions. Its depression is determined by the contraction of those which extend from its inferior border to the os hyoides. The contrary effect is produced by the action of the muscles, extending from the different points of its surface to the temples, and the neighboring parts of the head.[2] The power of the levator muscles of the jaw

Fig. 22.[1]

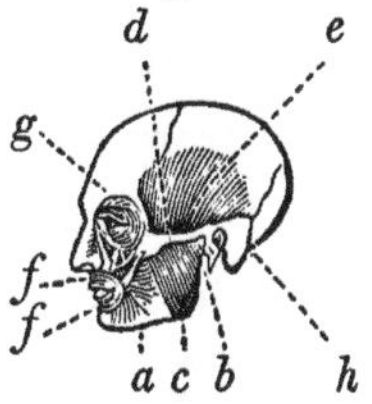

[1] *a,* Inferior jaw, — *b,* articulation of the lower jaw with the cranium, — *c,* masseter muscle, — *d,* zygomatic arch, — *e,* temporal muscle, — *f, f,* orbicular muscle of the lips, — *g,* orbicular muscle of the orbits, — *h,* occiput, or posterior part of the cranium.

[2] The principal levator muscles of the lower jaw, are, 1st, the *temporal muscle* (*e, fig.* 22), which springs from the coronoid process of this bone (see *fig.* 19, *e*), passes under the zygomatic arch (*d*), and extends upon the sides of the head to which it is fixed ; 2nd, the *masseter muscle* (*c*), which extends from the external face of the angle of the jaw to the zygomatic arch (*d*) ; 3d, the two *pterygoid muscles,* which occupy on the internal face of the jaw the place corresponding to that of the masseter, and are fixed to

is very great, and by their contraction, the substances introduced between the teeth are compressed the more forcibly, when placed near the bottom of the mouth, and consequently near the fixed points of these muscles.

The aliments are continually thrown between the teeth by the contractions of the cheeks, or by the motions of the tongue ; and thus pressed between two surfaces, hard, very unequal, and whose asperities are adapted to each other, these substances are soon divided into small portions, and crushed.

Influence of mastication upon digestion. The importance of this operation is very great, for the more complete the mastication the more easy is the digestion : which is as easily proved as understood. If an animal be made to swallow pieces of meat of various sizes, and after a certain time it is killed and the stomach opened, the smallest fragments will be the most advanced in digestion, and the superficies of the greater will hardly have been touched, while the smaller portions will be completely softened. Now this happens when fragments unequal in size, of any body susceptible of being dissolved in this liquid, are plunged into water, sugar, for example.

Insalivation of the aliments. While the aliments are submitted to the mechanical division, they imbibe saliva, and are sometimes even dissolved in it.

Saliva. The *saliva* is a colorless liquid, transparent, slightly viscous, which continually flows into the mouth, the lower parts of which it occupies. Chemical analysis has demonstrated, that it is composed of about 993 parts of water to the 1000 ; the other seven thousandths are

the base of the cranium on each side of the posterior opening of the nasal fossæ.

formed as follows ; of a peculiar animal matter about three thousandths, of mucus 1.4 ; chloride of sodium (or marine salt), chloride of potassium, tartrate of soda, and a small quantity of free soda, which gives to this liquid its alkaline properties, make up the remainder.

The mixture of the saliva with the aliments is a circumstance of more importance, than would at first be supposed. It facilitates mastication, aids powerfully deglutition, and, as we shall hereafter see, appears to perform an important office, in the digestion of these substances.

The glands, which form the saliva, are situated Salivary glands. around the mouth, and are composed of small agglomerated granulations. In man, there are three pairs, placed symmetrically upon each side of the head : the *parotid glands* situated in front of the ear, and behind the lower jaw ; the *submaxillary glands*, lodged under the angle of the jaw (*m, fig. 23*), and the *sublingual glands (l)* placed beneath the tongue, in the space between the two sides of the jaw.

Each of these glands communicates with the mouth by a peculiar excretory duct, and pours into it the saliva in variable quantities. With a tolerable appetite, the sight of aliments is sufficient to determine a more considerable afflux, and the presence of a foreign body in the mouth, even if it be completely insipid, always excites the secretion of this liquid : it would appear, that upon mastication it becomes more alkaline than usual.

So long as mastication is unfinished, the pos- Deglutition of the aliments. terior opening of the mouth is closed by the velum palati, which descends and rests upon the base of the tongue. The aliments cannot then penetrate farther into the digestive canal : but when this operation is terminated, this movable separation of the mouth from the pharynx is raised, and deglutition goes on.

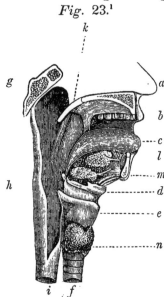

Fig. 23.[1]

This name is given to the passage of the aliments from the mouth to the stomach, through the pharynx and æsophagus.

The *pharynx*, or *posterior fauces*, is a cavity continuous with the mouth, and which is placed at the superior part of the neck (*fig.* 23 and 24.) By its summit it communicates with the nasal fossæ ; and above and in front, it is separated from the mouth only by the velum palati. Below and in front, the larynx (*e*), opens into it ; lastly,

[1] This figure represents a side view of a vertical section of the mouth and pharynx ; — *a*, the nose, — *b*, the upper lip, placed in front of the arch of the palate, which extends horizontally backwards, and separates the cavity of the mouth from the nasal fossæ, — *c*, the tongue, the base of which is fixed to the os hyoides (*d*), — *e*, the larynx suspended to the os hyoides, and opening into the pharynx, — *f*, portion of the trachea, a tube, which continues with the larynx at one end and at the other communicates with the lungs, — *g*, portion of the base of the cranium, to which is suspended the pharynx (*h*), — *i*, commencement of the æsophagus, — *k*, section of the velum palati ; above this partition is the posterior opening of the nasal fossæ, and below are two kinds of pillars, between which are situated the tonsils, — *l*, sublingual gland placed beneath the tongue, and communicating with the mouth by a small excretory duct, directed forward, — *m*, submaxillary gland, placed behind and below the preceding, — *n*, thyroid body, a kind of imperfect gland, placed in front of the inferior part of the larynx.

below and behind, it continues into the æsophagus (*i*);
a long and narrow tube, which descends the whole length
of the neck, traverses the thorax, passing between the
lungs, behind the heart, and in front of the vertebral
column, perforates the diaphragm, and at last terminates
at the stomach.

The *velum palati*, which
separates the mouth from
the pharynx, is a movable
partition, suspended transversely from the posterior
border of the palate, and free
at its inferior border, which
is prolonged in the middle
to a point, called the *uvula*
(*fig.* 23, *k*, and 24, *d*). It is
formed by a fold of the mucous membrane, which lines
the whole digestive canal,
and contains in its interior a

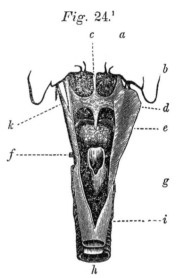

Fig. 24.[1]

[1] The pharynx seen from behind, and opened to display the relative position of the posterior openings of the nasal fossæ, of the velum palati,
base of the mouth and opening of the larynx; — *a*, base of the cranium, —
b, mastoid process of the temporal bone, situated at the side of the
base of the cranium behind the ear, — *c*, vertical partition of the two
nasal fossæ, the termination of which may be seen at the superior part of
the posterior fauces, — *d*, velum palati, making part of the arch of the
palate; in the middle of its inferior border is found a prolongation, called
the *uvula*, and on each side of this appendix is seen the buccal cavity, — *e*,
base of the tongue, — *f*, extremity of the os hyoides; on the opposite side
this bone is entirely concealed by the portion of the posterior wall of the
pharynx, which is turned outwards, — *g*, opening of the larynx or glottis;
conducting to the lungs by the trachea, a kind of valve, called *epiglottis*,
opens upwards and in front of this opening; it is here applied against the
base of the tongue, — *h*, portion of the trachea, — *i*, commencement of the
æsophagus, — *k*, one of the levator muscles of the pharynx.

great number of muscles, which allow it to execute many motions ; to descend and rest upon the base of the tongue, to be raised and carried obliquely backwards toward the posterior wall of the pharynx, so as to intercept more or less completely the passage between this cavity and the nasal fossæ.

Mechanism of deglutition.   *Deglutition* appears to be very simple, and yet it is really the most complicated of all the operations of digestion. It is produced by the contraction of a great number of muscles, and requires the concurrence of several important organs. All the muscles of the tongue, velum palati, pharynx, larynx, and æsophagus take part in it.

When it is to commence, the aliments are collected upon the back of the tongue, which is raised, and presses them from before backwards against the velum palati ; this partition then rises to a horizontal line, and thus permits the aliments to issue from the mouth ; if it did not oppose the motion impressed upon these substances by the movements of the tongue, the aliments would penetrate the nasal fossæ ; but the direction it occupies, obliges them to descend into the pharynx. This first period of deglutition is under the direction of the will ; but not so with the remainder of this operation, and the motions, by means of which the aliments arrive at the inferior part of the pharynx, are involuntary and in some sort convulsive. The alimentary bolus (for thus each mass of the aliment swallowed is called), then passes over but a very small space ; but it must avoid the opening of the larynx, and that of the nasal fossæ, where its presence would be injurious, while its passage must be so prompt as to offer but a momentary interruption to the free communication of the larynx with the external air.

Let us see how nature accomplishes this important result.

The alimentary bolus no sooner touches the pharynx, than every thing is called into action. This cavity contracts and embraces the alimentary bolus, while on the other hand the larynx ascends and passes in front of this body, to render more rapid its passage over the opening of the glottis. Finally, during this movement, the edges of the opening close exactly, and the epiglottis, pressed against the base of the tongue, descends so as to cover the entrance of the larynx. Thus the alimentary bolus, continually pressed by the contraction of the pharynx, slides upon the surface of the epiglottis, and arrives at the æsophagus, the circular fibres of which, by successive contractions, drive it into the stomach.

Fig. 18.

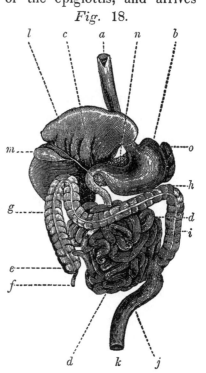

The *stomach* (*fig*. 18, *b*), is an enlarged portion of the alimentary canal, which is continuous with the æsophagus, and which is the seat of the most remarkable phenomena of digestion, the transformation of the aliments into chyme. It is a membranous sac placed across the superior part of the abdomen, and in form resembling a bag-pipe.[1] It

<hr />

[1] In fact, the bag-pipe is constructed from the stomach of those animals, in whom this organ most resembles man.

gradually diminishes from left to right, and curves upon itself, so that its superior border is concave and very short while its inferior border (called the *great curvature of the stomach*) is convex and very long. Toward either third of the stomach there exists, especially during digestion, a contraction which divides the organ into two parts ; that situated to the right called the *cardiac portion* of the stomach, that to the left called the *pyloric portion*. The opening, by which this viscus communicates with the æsophagus, is also called the *cardiac orifice*, because it is situated on the side of the heart. That which conducts from the stomach to the intestines is called the *pylorus*,[1] and is situated at the extremity of the pyloric part.

The walls of the stomach are very extensible : when its cavity is not filled with aliments they contract, and there are upon its internal face a multitude of folds, the number of which diminishes in proportion to the distention of the organ. We also remark upon the surface of the mucous membrane which lines the stomach, a very considerable number of small secretory cavities, called *gastric follicles*, which pour out upon the aliments the liquid they secrete.

Gastric juice. This liquid, which is called the *gastric juice*,

---

[1] The word pylorus is derived from the greek πυλουϱὸς, gate keeper, (πύλη a gate and οῦϱος a keeper), and has been given to the intestinal orifice of the stomach, to signify the functions it fulfils; while the aliments are as yet not sufficiently digested to allow of their passage into the intestine, the pylorus remains contracted and does not open to them a passage; but when the aliments are converted into chyme, this opening relapses and allows them to pass. The name of *valve of the pylorus* is given to a circular fold, which surrounds this opening, and which is formed by a fold of the mucous and muscular tunics of the stomach.

is, as we shall hereafter see, one of the most important agents of digestion, as by its action the aliments are converted into chyme. When the stomach is empty, it is formed in very small quantities ; but when the walls of this cavity are excited by the contact of food, and especially of solid food, the gastric juice flows in abundance, and has always very marked acid properties. This acidity appears to be due in part to the free hydrochloric acid, and partly to the presence of a peculiar substance, which is also met with in the milk, and is called the lactic acid. Some salts may also be detected in it, such as marine salt, phosphate of lime, etc., and about ninety-eight hundredths of water.

The alimentary substances, which accumulate Delay of the food in the stomach, are strongly pressed upon by stomach. the action of the muscular walls of the abdomen, and would remount into the æsophagus, if the portion of this duct near the cardia were not closed by the contraction of its muscular fibres. Sometimes this resistance is overcome, and the food reascends to the mouth, or is even thrown out, which bears the name of *regurgitation*, or *vomiting*.

On the other hand, the food cannot simply traverse the stomach, and then at once enter the intestines, for the opening of the pylorus is completely closed by the energetic contraction of the muscular fibres, by which it is surrounded. The food must therefore remain in the stomach, where it accumulates principally in the cardiac part, or great cul-de-sac of this organ. Some of the ingesta are there simply absorbed by the walls of the stomach, and penetrate into the blood without any previous alteration ; this is the case with water, diluted alcohol, and some other liquids. Other substances pass into the intestine, and are even expelled as excrements without any alteration ; but the aliments are here digested, and thus transformed into a pulpy and semi-liquid mass, called *chyme*.

Formation of the chyme. The fragments, placed towards the surface of the alimentary mass, and near the walls of the stomach, early imbibe the gastric juice, become acid as is this liquid, and gradually soften from the superficies towards the centre. The whole mass of the food finally undergoes the same alteration, and in consequence of this softening, these substances are transformed into a soft, pultaceous, grayish matter, of a faint and peculiar odor, which is chyme mixed with the remains of the food. A white substance is formed upon the walls of the stomach, resembling the white of an egg partially cooked, and which mingles with the other products of the stomachic digestion.

These alterations take place with more rapidity in the portions of the stomach near the pylorus than in the great cul-de-sac, and are propagated from the superficies of the alimentary mass to its centre.

Peristaltic movements of the stomach. While chymification is going on, the walls of the stomach become the seat of circular contractions, which at first proceed from right to left, so as to propel the chyme by which the alimentary mass is covered towards the great cul-de-sac of the stomach ; but after a certain time, all these vermicular motions, which are called *peristaltic*, take place in an opposite direction, and convey the chyme to the pylorus, and then into the small intestine.

All alimentary substances are not transformed Duration of the digestive process. into chyme with equal rapidity. The observations and experiments, which have been made upon this subject, demonstrate that muscular flesh is much easier of digestion than most herbaceous substances ; that cooking exerts a great influence upon it, and that boiled veal, for instance, is two thirds more digestible than roast ; that the skin and tendons resist a long time the action of the stomach, etc. Great differences however exist in different individuals : the size of the pieces swallowed also influences their transformation into chyme, as might easily be inferred from the nature of the digestive process.

In general, the aliments remain several hours in the stomach before they are completely transformed into chyme.

A great number of experiments have been Cause of the transformation of the aliments into chyme. made with the view of ascertaining what passes in the stomach during digestion. The most remarkable are those of Spallanzani, a celebrated physiologist of Modena. At the time when he entered upon his researches, it was thought that this phenomenon was only a species of trituration, and that the chyme was merely the food bruised till reduced to pulp : but Spallanzani showed this not to be the case. He caused birds to swallow articles of food contained in tubes, and little metallic boxes, the walls of which were pierced with holes, so as to preserve these substances from all friction, but not to withdraw them from the action of the liquids contained in the stomach, and he found that digestion took place as under ordinary circumstances. He therefore justly concluded, that the gastric juice must be the

17

principal cause of the chymification of food, and to be more completely assured, he had recourse to very ingenious experiments. He made crows and other birds swallow little sponges attached to a thread, by means of which he withdrew these bodies from the stomach after they had remained there for some minutes, and had imbibed the liquids contained in this cavity. Thus he procured a considerable quantity of the gastric juice, which he placed in small vessels with the food suitably divided ; he took care at the same time to raise the temperature, so as to imitate as nearly as possible the circumstances under which chymification takes place ; and at the end of some hours he found the alimentary mass, submitted to this artificial digestion, transformed into a pulpy matter, similar in all respects to that which would have been formed in the stomach by a natural digestion. Thereby proving that the action of the gastric juice upon the food is the principal cause of its transformation into chyme.

Intestines.   That portion of the alimentary canal, into which the food penetrates after its digestion in the stomach, bears the name of *intestine* (fig. 18, *c, d*). It is a membranous and convoluted tube, small in diameter, but very long, in man being about seven times the length of the body.

In animals nourished exclusively upon flesh, the intestines are, in general, shorter than in man, and other omniverous animals ; while in the herbivorous, their length is much more considerable. Thus in the lion it is only about three times that of the body, and in the ram it often equals twenty-eight times this length. The reason of this difference is obvious, for it is evident that herbaceous substances, which are very slow of digestion, and which contain but a really small portion of nutritive materials, must be taken in greater quantities, and must remain a longer time in the alimentary canal than muscnlar flesh, the digestion of which is very prompt, and of which nearly the whole mass is composed of nutritive materials.

The intestines, as we have already said, are lodged in the abdomen, and enclosed by the folds of the peritoneum, which fix them to the vertebral column. They are divided into two distinct parts, *the small* and *the large intestine.* The *small intestine* makes the Small intestine. continuation of the stomach, and in its interior digestion is finished. It is very narrow, and constitutes about three fourths of the whole length of the intestines. Its exterior surface is smooth ; the muscular fibres which surround it, are placed side by side ; and the mucous membrane, lining the interior, presents upon its surface many small *follicles,* and a great number of prominent appendages, called *villosities,* also a great number of transverse folds called *valvulæ conniventes.* The follicles secrete a viscous humor very considerable in quantity ; the villosities, as we shall afterwards see, appear to serve especially for the absorption of the products of digestion ; and the valvulæ conniventes to retard the course of the chyme.

Anatomists distinguish in the small intestine three parts, the *duodenum,* so called because its length is about equal to twelve fingers, the *jejunum,* because in the dead body usually found empty, and the *ileum,* (from ειλειν to turn, to twist), but this distinction is of little importance in physiology.

The alimentary matters which enter this intestine, are mixed with the humors secreted by Fluids contained in the small intestine.

its walls, and with two peculiar fluids, the *bile* and *pancreatic juice*, each formed in distinct glandular organs situated in the neighborhood of the stomach.

The liver. The liver (fig. 18, *l*,) which is the productive organ of the bile, is the largest viscus of the body. It is situated at the superior part of the abdomen, principally upon the right side, and descends to the lower border of the false ribs. Its superior face is convex, and its inferior irregularly concave. It is divided into three lobes, the largest of which is situated to the left, and separated from the right by a fissure, and the smallest (called the *lobule*) is placed under the others. The color of this organ is reddish brown upon the surface, and yellowish in the interior. Its substance is soft and compact, but traversed by a multitude of canals, and when torn it appears to be formed by the agglomeration of small solid granulations, into which the blood vessels empty, and from which originate the excretory ducts carrying off the bile.

These excretory ducts successively unite to form twigs, branches, and lastly a trunk, which issues from the inferior face of the liver in the direction of the duodenum; and which, in its course communicates with a membranous sac adherent to the liver, habitually distended with bile, and called the *gall-bladder*. The termination of this canal may be seen in the duodenum, at a short distance from the stomach.[1]

The liver presents a very remarkable peculiarity. The

[1] The excretory duct immediately from the liver is called the *hepatic duct*, and that from the bladder the *cystic duct*. Finally the common trunk, formed by the union of these two vessels, is called the *ductus choledochus* (from χολη, bile and δοχός, the container.

greater part of the blood, which circulates in this organ, is not arterial, as in the other parts of the body. The venous blood coming from the intestines enters it by the vena porta, which ramifies like an artery, and it would even appear, that the formation of the bile depends principally upon this liquid.

The *bile* is a viscous, ropy liquid, greenish, and   Bile. of a very bitter taste. Its chemical composition is very complicated, for we find in it water, albumen, resinous matter, a yellow coloring principle, various fatty matters, several salts, and free soda. It is always alkaline, and has some analogy to soap.

The bile is constantly flowing into the intestine, but the flow appears to be increased in quantity during digestion ; for when the stomach is empty the gall-bladder is full, and when digestion is terminated, this reservoir is nearly empty.

The *pancreatic juice* is very analogous to the   Pancreatic juice. saliva, both in physical properties and chemical composition; the *pancreas*,[1] which forms it, resembles the salivary glands. It is a granular mass, divided   Pancreas. into a great number of lobes and lobules, of firm consistence, whitish gray color bordering upon red, placed crosswise between the stomach and the vertebral column (fig. 18, *n*). From each of its granulations there originates an excretory duct, all of which unite together as the veins, and thus form a canal opening into the duodenum near the mouth of the ductus choledochus.

We have already seen how the peristaltic move-   Delay of the chyme in the ments of the stomach propel the chyme into the   intestine.

---

[1] The word *pancreas* signifies all of flesh (from παν, all, and κρèaς, flesh), and was given to this gland by the ancients.

duodenum through the pylorus.   This opening is supplied
with a valve, which opposes the return of this material
to the stomach, and the presence of the chyme in this
intestine causes in it contractions analogous to those of
the stomach, and which exactly resemble the movements
of a crawling worm.   By the aid of these movements
the chyme accumulates in the intestine, and gradually
advances farther in the tube.   During this passage it
mixes with the bile, and the other humors in its course,
and gradually changes its properties ; it becomes yellowish, bitter, loses its acidity, then becomes alkaline, and
at the same time there is separated from it a matter more
or less thick, sometimes white, sometimes gray, according
to the nature of the aliments, from which it arises, which
Chyle.   is attached to the surface of the intestinal mucous
membrane, and which bears the name of *chyle*.   This
matter, as we shall afterward see, is absorbed, and toward
the inferior third of the small intestine is no longer met
with ; the paste formed by the residue of the chyme, by
the bile, and the other humors already mentioned, acquires in this portion of the alimentary tube more consistence, takes a deeper color, and passes into the intestine to be rejected as excrement.

Intestinal
gasses.   Digestion then is finished in the small intestine,
and during its performance various gasses are disengaged
from the alimentary mass, which more or less dilate the
intestine.   These gasses are principally carbonic acid
and pure hydrogen ; sometimes too azote is found with
them.

Large intes-
tine.   The *large intestine*, (*fig.* 18, *e, g, h, i,*) which
is the continuation of the small, and which receives the
residue left by digestion, may be easily distinguished by

the numerous dilatations of its walls between the collections of its muscular fibres. It is divided into the *cœcum, colon* and *rectum*. The *cœcum*,[1] which is   Cœcum.
situated near the haunch bone of the right side, is prolonged into a cul-de-sac beyond the point of insertion of the small intestine, and presents at this extremity a vermiform appendix. Folds arranged as valves supply the opening from the small intestine, and prevent the matters driven into the cœcum from reëntering the ileon, and returning to the stomach.

The colon, (derived from κωλύω, I stop, because   Colon.
this intestine retains for a long time the excrementitial matters in its folds), is continuous with the cœcum, remounts towards the liver, traverses the abdomen directly beneath the stomach, and redescends on the left side to gain the pelvis, where it continues into the rectum, which terminates at the anus.

The residue, arising from the digestion of the   Progress of the remainder of digestion in the large intestine.
food, is driven by degrees from the cœcum to
the rectum, so called because straight, where it
accumulates, and remains for a longer or shorter period. By thus traversing the large intestine these matters acquire greater consistence, and a peculiar odor. There is developed at the same time in this intestine a considerable quantity of gas, differing essentially from the gases of the small intestines by the almost constant presence of carbonated hydrogen, and sometimes also by the presence of a little sulphureted hydrogen.

The fleshy fibres around the anus, and which form the

---

[1] Anatomists have named the first portion of the large intestine *the cœcum*, because it is prolonged inferiorly in the form of a cul-de-sac (from *cœcus*, blind).

sphincter muscle of this opening, are continually contracted, and consequently oppose the exit of matters ac-
cumulated in the large intestine. In general, the contraction of the muscular fibres around the intestine does
not suffice for the expulsion of their contents, the diaphragm and the other muscles of the abdomen concur to
the same end, by compressing the mass of the viscera
contained in this cavity.

Theory of
digestion.   Such are the principal phenomena of digestion.
Let us now inquire, if, in the present state of science, it
is possible to explain in a satisfactory manner the different changes experienced by the food during the performance of this function.

The experiments of Spallanzani and of some other
physiologists show, that the principal agents of digestion
are the various liquids, which moisten the food in the
different parts of the digestive canal. These juices are
of three kinds : 1st, the saliva, always alkaline : 2nd,
the gastric juice, which is acid : 3d, the bile and pancreatic juice, which are alkaline as the saliva.

By the action of the saliva the food is sometimes dissolved, in most cases, however simply softened, and often
without much change in its physical properties. It
would appear that this liquid plays an important part in
digestion, as may be seen from what takes place in ru-
minating animals.

In these animals there are four distinct cavities, to
discharge the functions of the single stomach of man.
The food is at first introduced into a large sac called the
paunch, when it remains a certain time, and then passes
into the second stomach (*the reticule*), and is thrown up
from the latter into the mouth, to be bruised by the teeth

and moistened with saliva; it next descends into the *many-plies* or third stomach, and thence into the *reed.* The experiments of Prevost and Leroyer of Geneva show, that the aliments contained in the paunch and the reticule are moistened with an alkaline juice, and that by its action the albumen and some other substances, of which they are in part composed, are dissolved. If by pressure this liquid is forced out, and acid thrown upon it, there is immediately formed a flaky precipitate, similar to the white of an egg half cooked. Now precisely the same takes place, when the alimentary mass passes into the many-plies: it meets there an acid juice, and deposits upon its walls a whitish layer, which is nothing but chyme.

On the other hand the experiments of Spallanzani, which we have already had occasion to mention, show that the food may also be directly attacked by the gastric juice. This liquid may dissolve substances, on which the saliva is inert, and the principles thus dissolved must in their turn be precipitated as solid globules, when the alkaline juices contained in the small intestine are mingled with the acid products of the stomachic digestion.

Digestion seems then to be the result of the chemical action of the saliva, gastric juice, and bile, upon the aliments, and upon the materials extracted from these substances by the action of the digestive liquid, to which they have been submitted before meeting either of these two latter agents. This phenomenon would then essentially consist in the solution of the alimentary matters, and their subsequent precipitation in the globular state; but it must be confessed, there are very many points yet

18

to be elucidated relative to the theory of digestion, and this question, the interest and importance of which every one can appreciate, demands a new investigation.

Absorption of the chyle.   To terminate the study of this function, it remains for us to examine how the nutritious matter, extracted from the food by the digestive function, can pass from the intestinal canal into the mass of the blood, which it is destined to renew.

Chyliferous vessels.   Some of the liquids introduced into the stomach are directly absorbed by the veins in the walls of this cavity, and in those of the small intestine; but the chyle follows another route, and penetrates into a particular set of canals, destined to effect its transport. These vessels called *chyliferous* (or *lacteals*, from the appearance they take when filled with chyle) belong, as we have already said, to the lymphatic system. They arise from imperceptible orifices on the surface of the villosities of the intestinal mucous membrane, and unite like veins into larger

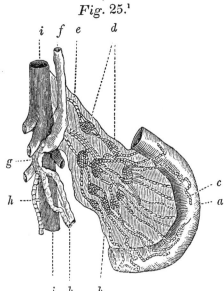

Fig. 25.[1]

1 A portion of the small intestine with the chyliferous vessels originating in it and the commencement of the thoracic canal.

a, portion of the intestine, — b, mysentery which fixes the intestine to the posterior wall of the abdomen, — c, radicles of the chyliferous vessels or smaller branches, which proceed between the two folds of the mysentery to the vertebral column. During this course the lymphatic vessels traverse a number of small bodies, irregular in form, and of a pale rose color, which are called the *mesenteric glands* (*d*), and after issuing from these glands they unite into a common trunk, called the *thoracic duct* (*f*). This canal also receives the lymphatic vessels from almost all other parts of the body. It traverses the diaphragm, and mounts in front of the vertebral column to the base of the neck, where it finally terminates in the left subclavian vein. There are in its interior, folds similar to the valves of the veins, so arranged, as to permit the passage of the liquids towards the subclavian vein, but to prevent their return to the intestine.

If an animal is fasting, these vessels are nearly empty, but when the intestinal digestion is in full activity, they are soon gorged with chyle.

The physical properties of this liquid vary with Chyle. the nature of the food, from which it arises, and the animals in which it is observed. In man and most mammiferæ, the chyle is usually a white liquid, opaque, having nearly the aspect of milk, of a salt alkaline taste, and of a peculiar odor. Examined by the microscope, it presents a multitude of globules, analogous to those forming the central nucleus of the globules of the blood. When at rest, it soon forms a mass, like the blood, and after the

creeping upon the intestine, — *d*, mesenteric glands, — *e*, chyliferous vessels after their passage across the mesenteric glands, — *f*, thoracic duct, — *g*, enlarged portion of the thoracic duct called the reservoir of Pecquet (better known as the receptaculum chyli), — *h*, *h*, lymphatic vessels of the inferior limbs going to the thoracic duct, — *i*, *i*, portion of the aorta, by the side of which the thoracic duct ascends to gain the subclavian vein.

expiration of some time it separates into three parts; a solid clot, which occupies the bottom of the vessel, a liquid analogous to the serum, and a very thin layer, which swims upon it, and which appears to be of a fatty nature. The chyle also takes during coagulation, a lively rosaceous tint, and if agitated with oxygen the phenomenon is yet more marked.

Chyle, arising from articles of diet, which do not contain fatty substances, is much less opaque than that furnished by matters containing fat or oil, and the layer which forms upon the surface after coagulation is much less thick. The solid clot, which is principally composed of fibrine and coloring matter, is very small in quantity in the chyle arising from the digestion of sugar, gum, etc., while from the chyle furnished by muscular flesh is formed a considerably large one.

Mechanism of the chylous absorption. The villosities, with which the surface of the mucous membrane of the intestines is supplied, appear to be the special agents of the absorption of the chyle. As soon as it commences, we find them swollen and saturated with this liquid, as sponges which have imbibed milk; some anatomists have imagined they could distinguish in this species of fringe, very small openings communicating with the radicles of the lymphatic vessels; and if this be the case, we can easily comprehend, how the chyle may penetrate into these vessels without being absorbed by the veins. This liquid contains, as we have already said, globules too large to pass through the simple porosities of the venous walls, while they would find an easy access into the chyliferous vessels through the holes, with which the villosities appear to be pierced.

However, the chyle penetrates into these latter vessels, and flows with considerable rapidity the length of the thoracic duct to the left subclavian vein; if this canal be tied in the living animal, the passage of the chyle into the circulatory system is completely cut off, and this liquid accumulates in the thoracic duct. The cause of its movement of ascension in this canal, and in the numerous chyliferous vessels, which represent the roots of this trunk, is not known. It is found to continue some time after death, and it has also been ascertained, that the course of the chyle is favored by the respiratory movements, the beating of the arteries, and all the movements, which compress in an intermitting manner the thoracic duct, and which can be readily comprehended, by reason of the valves already mentioned, and the action of which has been explained when treating of the venous circulation.

The chyle thus mingling with the blood serves <small>Uses of the chyle.</small> to repair the losses experienced by this liquid in its action upon the organs nourished by it. But how is this hœmatosis, or transformation of chyle into blood, effected?

We have already said, that these two liquids are much alike, and that by the action of the air upon the chyle, this resemblance becomes yet closer, because the color of this liquid is rendered very analogous to that of the blood. From which it might be concluded, that a part of the modifications necessary to change the chyle into blood, take place in the interior part of the lungs, and by the act of respiration.

Still there exists between these two liquids an important difference, not to be thus accounted for. The globules of the chyle do not appear to be contained in a

colored vesicle, like those of the blood, and therefore
the question arises, where are these latter globules
formed ?

This question is not yet completely determined ; but
from some acknowledged facts, the liver would seem to
be the agent in effecting this important change.

URINARY SECRETION.

One portion of the foreign substances absorbed by the
human body and of the materials eliminated from the organs by the exercise of nutrition, are expelled from the
economy, either by the respiration, pulmonary exhalation,
and cutaneous transpiration, or by the secretion from the
surface of the intestines and the other mucous membranes ; but substances, which are useless or injurious to
the economy, may be rejected by another way, the urinary excretion.

Urinary ap-   This function is that of the *kidneys*, which are
paratus.
two voluminous glands situated in the abdomen, on either
side of the vertebral column, between the muscles of the
lumbar region of the back and the peritoneum, and ordinarily surrounded by much fat ; their color is reddish
brown, and their form like that of a hancot or kidney-
bean. The panenchyma appears to be formed of two
substances ; one superficial, called the cortical or glandular ; the other interior, named tubular or mamellated.
The cortical substance is formed of extremely small
granulations, and a multitude of capillary canals twisted

upon themselves, and united in clusters; the mamellated is composed of canals, which spring from the cortical, and, converging towards the middle of the interior border of the gland, form by their union a certain number of cones, with a rounded base, surrounded by the cortical layer. These canals all empty at the summit of these pyramids into other and greater ducts called the *calices*, which in their turn supply the *pelvis*, a small membraneous pouch situated in the fissure of the internal border of the kidneys.

These glands receive a considerable quantity of blood from a large artery, which ramifies upon the cortical substance, where is effected the secretion of a peculiar liquid, the urine.

This liquid descends by the canals composing the mamellated substance, and by the calices to the pelvis, and thence passes into the bladder through a long membraneous tube of the size of a writing quill, which passes transversely from the pelvis to the bladder, and is called the *ureter*. The *bladder* is a conical pouch, which fulfils the functions of a reservoir for the urine, and is situated at the inferior part of the abdomen, behind the anterior portion of the pelvis, called the arch of the pubis. It is formed of a mucous membrane, surrounded by fleshy fibres, and continues inferiorly with a narrow canal, opening outwards, and which is called the *urethra*.

The urine is a yellowish and acid liquid, which,   Urine. in man, is composed in the normal state of nearly ninety-three hundredths of water, three hundredths of a peculiar matter, called urea, of a thousandth of uric and a small quantity of lactic acid, and of several salts (the chloride of sodium or marine salt, phosphate of lime, &c.) In

the carnivorous mammiferæ its chemical composition is
nearly the same as in man, except that we do not find
the uric acid ; but in young children and herbivorous animals, there is a very peculiar substance contained in it,
namely, the hypo-uric acid ; and in birds, as well as in
most reptiles, (lizards, serpents, &c.) it scarcely contains
any thing but uric acid ; finally in frogs and turtles, we
find both urea and albumen : its composition appears to
be nearly the same in fishes, but in insects there is uric
acid.   In certain diseases, its composition, however, is
considerably changed.

Source of the   The rapidity, with which fluids introduced
  urine.
into the stomach pass into the bladder and are expelled
by the urinary passages, is very great.   Every one has
made the remark, and the experiments upon living animals also prove it.   But yet there exists no direct communication between these two organs, and the liquids
can only pass to the bladder after having been absorbed,
mingled with the mass of the blood, thus conveyed to
the substance of the kidneys, and separated by the secretory exercise, of which these glands are the seat.

It is evidently from the blood that the kidneys derive
all the aqueous portion of the urine ; and if certain substances, easily recognised, (such as rhubarb, or the yellow cyanuret of potassium or iron,) be introduced into
the current of the circulation, (either by injection or absorption,) they are soon found to be expelled in the
urine.

The blood, then, furnishes to the kidneys the materials
necessary to form the urine ; and the knowledge of this
fact must naturally lead physiologists to inquire if the
several principles contained in the latter existed, com-

pletely formed, in the blood, and were merely separated by the action of the kidneys, or whether these organs produced them, by their action upon other substances contained in the blood.

Water and most of the matters expelled by the urinary passages, exist in quantities more or less appreciable in the blood ; but, under ordinary circumstances, chemical analysis does not reveal to us the presence of urea and the other principles essentially characterizing the urinary secretion. It might therefore be supposed, that these materials were formed directly by the kidneys ; but it is not so ; these organs only separate them from the blood gradually, as they appear ; and an infallible proof will be found in the extraction of the kidneys in the living animal ; for then, the urinary secretion being interrupted, we find urea in the blood.

Therefore it is a legitimate inference, that the urinary glands actually derive from this liquid the substances which compose the urine, and that they find them ready formed.

But different circumstances influence the ac- Circumstances influencing the tivity of this function, and may modify both the activity of the secretion. mass of the liquids expelled by the urinary passages, and the quantity of solid materials, separated from the blood by the action of the kidneys, and held in solution by the aqueous part of the urine.

The quantity of water expelled by the urinary secretion depends, in a great measure, upon that of the fluids taken into the stomach.

Water introduced into the mass of the blood by absorption separates from it more or less rapidly, so that after a certain time the equilibrium is established in the

19

economy, however great the quantity of fluids introduced into the stomach. This liquid escapes from our bodies in two distinct ways, namely, by pulmonary and cutaneous exhalation, and by the urinary secretion; and these two functions are in a measure mutually dependent, and the mass of fluids in circulation remaining the same, every thing which tends to diminish the one, tends to augment the other.

For example, the action of heat upon the body tends to augment the transpiration, and, consequently, to diminish the urinary secretion: thus the latter function is more active in winter, than in summer,[1] and if any one take a considerable quantity of drink, he can almost voluntarily determine its expulsion, by one or other of these means, according as he is placed in circumstances favorable for transpiration, or the urinary secretion.

The quantity of solid substance expelled by the kidneys, and held in solution by the aqueous portion of the urine, depends, in a great measure, upon the nature and abundance of the aliments taken.

It has been proved by M. Chossat, that when nourished upon the same aliments, the only variation being in quantity, the secretion of urea and the other different principles, with the exception of water, expelled by the kidneys, varies in the same proportion. It diminishes with rigorous abstinence, and augments with the increase in the quantity of food, always provided this quantity be not too great for digestion.

---

[1] The curious experiments of M. Chossat demonstrate, that in the cold season the mass of urine surpasses that of the fluids taken into the stomach. In the months of spring, when the temperature is mild, this relation is sensibly diminished, and in the hot season the proportion of the urine to the drink is only about nine tenths.

This physiologist has also proved, that the secretion of these matters increases in the ratio of the quantity of animalized substances taken, that is, substances which contain a considerable proportion of azote. Thus by nourishing himself upon bread alone, or upon flesh alone, he found, that for an equal weight of food (abstracting the water contained), the quantity of solid principles expelled under the form of urine was four times as great in the latter case, as in the former.

If we compare the quantity of carbon, azote, hydrogen, and oxygen, which enter into the composition of the aliments employed by man, with that of the same elements expelled under various forms, either by the lungs and skin, or by the kidneys, we find almost all the carbon, thus introduced into the body, escapes from the lungs under the form of carbonic acid, while the azote is almost entirely expelled by the urine, as urea, uric acid, &c.

But the state of the animal's system also exercises great influence upon the results of the urinary secretion ; as every thing which tends to weaken, appears to retard this secretion, and to diminish the exhalation of carbonic acid by the respiration.

The urine sometimes deposits in the interior _{Gravel and urinary calculi.} of the urinary passages various substances, which may be held in solution, but which when deposited as solids are styled gravel and urinary calculi.

The gravel is almost always formed of uric acid, and depends upon its too great secretion ; thus this malady is increased by every thing which tends to augment the proportion of solid substances held in solution by the urine, as the diet of the animal, the too constrained use of aqueous drinks, &c.   In general this deposit forms in the kidneys and is passed out with the urine.

The urinary calculi are larger concretions, which also sometimes form in the kidneys, but more generally are developed in the bladder where they reside.   They gradually enlarge by the addition of new matter from the urine, and owing to their formation present concentric layers, more or less distinct.

The substances entering into their formation are very various.   Some are always existent in the urine, but in quantities so small, that they are ordinarily held in solution.   Others are produced, or rendered insoluble by the chemical action experienced by the urine when long exposed to the air, or confined in the bladder.   Finally, others are the result of an abnormal action of the secretory organ itself.

The first are of uric acid, the second urate of ammonia, the phosphate ammoniaco-magnesian, the phosphate of lime, the third the oxalate of lime, the cystic oxide, etc.

Calculi of the first class are the most common, and it often happens that their presence in the bladder determines the deposition of the salts, that we have ranged in the second category.   It is rare to see these latter substances form the nucleus of a calculus ; but nothing is more common, than to find a nucleus of uric acid or oxalate of lime encrusted with earthy phosphate.

REVIEW.

Having now gone over the different series of phenomena, by means of which the nutrition of the body of animals is effected, to embrace at a glance the operation of all these functions, it is thought proper to give their enumeration in an order varying from that adopted in their study.

It has been shown, that all living beings must continually draw to the interior of their bodies water, oxygen, and various other alimentary matters derived from the exterior world, and deposit these new materials in the tissue of their organs.

The name of *absorption* has been given to this passage from without inwards, and to the combination of the materials, thus sucked in by the living organs with the mass of the nutritive fluid.

In plants, all nutritive substances are absorbed at once, and penetrate directly into the parenchyma of the organs, without having undergone any previous preparation. In animals, certain substances, as water and the oxygen of the air, are absorbed in the same manner, either by the skin, or by the internal tegumentary surface lining the aërial and digestive passages; but with most nutritive materials, it is quite otherwise, and the food cannot serve for the support of the body, and penetrate into the interior of the organs, until it has first been transformed into a peculiar liquid, called *chyle*, a transformation, which constitutes the phenomena of *digestion*.

The chyle, absorbed by the lymphatic vessels, mingles with the blood, and furnishes the materials of which this liquid is composed.

The blood circulates in all parts, and conveys to them the materials necessary for their support. By the aid of the nutritive principles, which are furnished them by this liquid, the living tissues continually incorporate with their own substance new particles, and while this exercise of assimilation is going on, they abandon other molecules, which entered into their composition, and the renewal of which has therefore become necessary.

This continual movement of the composition and decomposition of the solid parts of the body, constitutes *the exercise of the function of nutrition.* When the quantity of foreign materials, thus assimilated to the substance of the organs, exceeds that of the matters eliminated from these same organs, the body increases; in the opposite case, it becomes lean; and if these two phenomena are equally active, the weight of the animal remains stationary, although the materials of which its body is composed are unceasingly renewed.

The excrementitial matters, separated from the substance of the living organs by the exercise of nutrition, must be thrown out, — and this actually takes place. They are at first mingled with the blood, which conveys them along with it far from the parts where they were detached, and in its turn this liquid is freed from them, either by the simple exhalation which takes place from all tegumentary surfaces, exterior as well as interior, or by the secretion from the glands, the parenchyma of which it traverses.

The water thus expelled from the body, escapes principally by the insensible transpiration of the lungs and of the skin, and by the urinary secretion.

The greater part of the azoted principles and of the materials not volatile, are in general excreted by the kidneys.

On the contrary, from the lungs are exhaled carbonic acid and the other volatile principles which may be found mingled with the blood.

Respiration is consequently, at the same time a function of assimilation and of excretion, as it serves as a means for the entrance of the oxygen necessary to the support of life, and for the exhalation of the carbonic acid produced by the nutritive decomposition of the organs.

With regard to the nature of the movement of nutrition, of which all living parts are the seat, we again repeat, nothing certain is known: we can only presume, that it must resemble the secretory exercise, and that it must depend upon the action of an analogous cause, which cause appears to be the nervous influence, and which seems to have a great analogy to the physical force which produces the electro-chemical phenomena. Recent experiments by M. Becquerel upon the influence of electricity on the vegetation of plants, come to the support of this opinion.

FUNCTIONS OF RELATION.

In the enumeration of the different faculties, with which animals are endowed, we found that some were exclusively destined to confirm the existence of these beings, while others served to convey to them a knowledge of surrounding objects. The former constitute the functions of nutrition, which we have already studied; the latter, the functions of relation, with which we are now to be occupied.

If we examine what takes place in an animal Contractility. of the simplest structure and most limited powers, the first thing we notice is its motion, and that the movements which it executes are determined and directed by a cause acting internally. Among these movements, some are repeated in the same manner, whatever be the circumstances in which the animal is placed, and over which it has no actual control.

Volition. But there are others, which vary according to the wants of the animal, and are under the direction of an intelligent power, known as *volition.*

These two orders of phenomena constitute two of the most important functions of relation, to wit : contractility or the faculty of executing spontaneous movements; and volition, or the faculty of exciting contractility, and varying its effects with a view of arriving at a result foreseen by the animal.

Sensation. There is yet another property inherent in all animated beings, and which is still more remarkable ; it is sensation, or the faculty of receiving impressions from external objects, and being conscious of them.

Instinct. These three faculties are common to all animals, but they are not the only ones observed. It is remarked, that in all there exists an inward power, which causes them to perform certain acts useful for their preservation, but the result of which they cannot certainly presage, and the cause of which depends upon no apparent need. Thus a multitude of animals construct with the most admirable art dwellings, destined to lodge their offspring, and calculated to supply all their wants, and they always do this in the same manner, and with the same skill, even when they have been separated from others of their kind from birth, and have never seen analogous labors. Others at a determinate season of the year emigrate to far countries, where the climate will be more favorable to them, and they direct themselves thitherward as certainly, as if the end of their voyage were before their eyes.

The name *instinct* has been given to the cause, which thus constrains animals to execute certain determinate acts, which are neither the effect of imitation, nor yet the result of reasoning. These inclinations vary, so to speak, in every species of animals, and the phenomena, which result from them, are sometimes extremely simple, and sometimes of an astonishing complication.

Other animals more highly privileged, enjoy Intelligence. the possession of the *intellectual faculties*, or the power of recalling to the mind the ideas primarily produced by the sensations, of comparing them, of drawing from them general ideas and deducing motives of conduct.

Finally, there are also some animated beings, Expression. which enjoy the faculty of imparting their ideas to their fellows, either by certain movements, or by the production of various sounds.

The various phenomena, by which animals are put in relation with surrounding objects, may therefore, as we have seen, be referred to six principal faculties ; *sensation, contractility, volition, instinct, intelligence,* and *expression.* The four first exist in all animals, the two latter only in a small number, and the manner in which they are executed varies almost to infinity.

In some animals of very simple structure, the polypi for instance, the various faculties of relative life are not the appropriation of any particular organ : each separate portion can feel and move without the concurrence of another portion ; but in man and the immense majority of animals, the exercise of all these functions is dependent

20

upon the action of a determinate part of the body, which is called the nervous system.

Nervous tissues. This system is formed of a peculiar substance, soft and pulpy, which is almost fluid in the early periods of life, but which acquires consistence as man approaches the adult age.

It is worthy of observation, that in this respect, the inferior animals resemble more perfect beings whose development is not completed. In frogs the cerebral pulp affords no more consistence than in the human fœtus, and in crabs it is almost liquid. There is in nature a tendency of which we shall often have occasion to speak, viz., that of causing the superior order of animals to pass through transitory stages, analogous to the state which is permanent for beings of a less perfect structure.

The aspect of this substance, which is named the *nervous tissue*, varies greatly; sometimes it is white, at others gray and ashy; sometimes it is gathered into considerable masses, at others it forms long ramified cords. The latter organs are named *nerves*, and the former *ganglions* or *nervous centres*, for they serve as a point of union for all the filaments in question.

In man, and all the animals nearly allied to him, the nervous apparatus is composed of two parts, called *nervous system of animal life* or *cerebro-spinal*, and the *nervous system of organic life* or *ganglionary*; and each of these systems, in its turn, is composed of two parts; one central, formed of nervous masses more or less considerable, the other a periphery, formed of nerves going from these centres to various parts of the body.

Encephalon. The central portion of the cerebro-spinal nervous system is often designated as the cerebro-spinal axis, or the *encephalon*. It is essentially composed of the brain, cerebellum and spinal marrow, and it is lodged in a bony sheath formed by the cranium and vertebral column, or spine.

The cavity of the cranium occupies all the su- Parts protecting the encephalon. perior and posterior part of the head. It is of an oval form, and presents, on its inferior surface, a great number of holes, which give passage to the nerves passing out, and to the blood-vessels which serve for the nutrition of the parts contained in its interior. Lastly, in the point where the head articulates with the vertebral column on which it rests, there exists a great opening called the occipital hole, through which the cavity of the cranium is continuous with a canal, which extends the whole length of the median line of the back.

This canal is formed by a succession of bony Vertebral column. rings, called *vertebræ* (fig. 26), which joined *Fig. 26.*[1] together in a solid manner, constitute a kind of stalk, which occupies the whole length of the body, and is called the *spinal* or *vertebral column* (fig. 27). On each side there exists a *Fig. 27.* series of holes, through which the nerves pass to the different parts of the body.

Several membranes also surround the encephalou, and serve to fix or to protect this organ, whose structure is very delicate, and importance very great.

The first of these tunics is called the Dura mater. *dura mater;* it is a fibrous membrane, firm, thick, whitish and moist, which adheres by several points of its exterior surface, to the walls of the

---

[1] Fig. 26. The superior surface of one of the vertebræ of the back. Fig. 27, a profile View of the vertebral column.

cranium and vertebral canal, and forms around the nervous system a very resisting sheath. Upon its interior surface are many folds, which bury themselves in furrows, more or less deep, of the encephalic nervous mass, and form a kind of division, which prevents these parts from being displaced, and sustains them so that they do not press against each other in any position of the body. There also exist between its internal and external surface, very large venous canals, which are called *sinuses of the dura mater*, and which serve as reservoirs for the blood coming from the different parts of the encephalon.

Arachnoid. Within the dura mater we find a second covering, the *arachnoid*, so called from its tenacity and transparence, which have caused it to be compared to a spider's web. It belongs to the class of serous membranes, and represents a sac without an opening doubled upon itself, which envelops the encephalon, and lines the walls of the cavity of the dura mater, in the same way that the pleura envelops the lungs, and the peritoneum the intestines. Its interior surface, every where in contact with itself, is lubricated by a serous humor, and its internal plate penetrates into the different cavities, of which we shall hereafter have occasion to speak, in the interior of the brain. Its principal use is to furnish a liquid which bathes this organ and facilitates its movements.

Pia mater. Finally, we find beneath the arachnoid a third cellular membrane, which is wanting in certain parts, and which is called the *pia mater*. It is not a membrane, properly so called, but a cellular layer destitute of consistence, in which are ramified and interlaced, in a thousand different directions, a multitude of blood vessels, more or less fine and tortuous, which come from the encephalon, or are scattered in its substance ; the circulation of the blood in the encephalon being in a very peculiar manner. The arteries before penetrating into this organ, the texture of which is very delicate, are reduced to capillary vessels, the purpose of which is to moderate the force with which the blood reaches the organ.

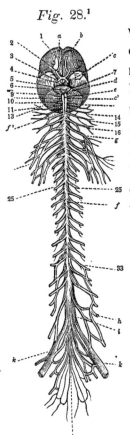

Fig. 28.[1]

The cerebro-spinal axis, Encephalon. which is protected by these different envelopes, is composed, as before mentioned, of several distinct organs ; but all these parts are intimately united together, and may be considered as a continuation of each other. Its anterior or superior part is very voluminous, and occupies the interior of the cranium : to this belongs the special name of *encephalon*. It is divided into two parts, the brain (cerebrūm), and the cerebellum, both continuous inferiorly with a large nervous cord lodged in the spinal column, and called the *spinal marrow*.

The *brain* (fig. 28, *a*, and fig. 29, *A*, *B*, *C*,) is the most voluminous portion of the encephalon ; it occupies all the superior part of the cranium, from the forehead to the occiput. Its form is oval, with the large

---

[1] Cerebro-spinal system seen on its anterior face (the nerves being divided at a short distance from their origin), — *a*, brain, — *b*, anterior lobe of the left hemisphere of the brain, — *c*, median lobe, — *d*, posterior lobe, almost

extremity behind, its superior face is quite a regular arch,
and at the sides it is a little compressed, while below it

Fig. 29.[1]

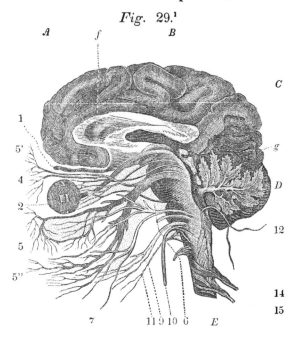

entirely concealed by the cerebellum, — *e*, cerebellum, — *f'*, medulla oblongata, — *f*, spinal marrow, — 1, nerves of the first pair, or olfactory, —
2, optic, or nerves of the second pair, — 3, nerves of the third pair, which
originate from behind the interlacing of the optic nerves in front of the
pons varolii, and above the peduncles of the brain, — 4, nerves of the fourth
pair, — 5, trifacial or nerves of the fifth pair, — 6, nerves of the sixth pair
lying upon the pons varolii, — 7, facial or nerves of the seventh pair, and
acoustic or nerves of eighth pair, — 9, glosso-pharyngeal, or nerves of the
ninth pair, — 10, pneumogastric, or nerves of the tenth pair, — 11, nerves
of the eleventh and twelfth pairs, — 13, suboccipital, or nerves of the thir-
teenth pair, — 14, 15, 16, three first pairs of cervical nerves, — *g*, cervical
nerves forming the brachial plexus, — 25, a pair of the dorsal nerves, — 33,
a pair of the lumbar nerves, — *h*, lumbar and sacral nerves forming the
plexus, from which arise the nerves of the lower extremities, — *i*, and *j*,
termination of the spinal marrow called the *cauda equina*, — *k*, great sciatic
nerve going to the lower extremity.

[1] Vertical section of the cerebrum, cerebellum, and medulla oblongata ; —

is somewhat flattened. At first, we distinguish two lateral halves, named *hemispheres of the brain,* and separated by a deep fissure, in which is buried a partition, formed by a fold of the dura mater, and called from its form the cerebral falx. From before backwards, this fissure divides the brain in its whole extent; it does not' however extend completely through from above downward, but is bounded inferiorly by a medullary plate, which extends from one hemisphere to the other, and is called the *corpus callosum* (fig. 29, *f*). The surface of the hemisphere is divided by a great number of tortuous and irregular furrows of various depths, which separate the rounded eminences upon its exterior, convoluted upon themselves, and having some resemblance to the folds of the small intestine, in the abdomen. These eminences are called the convolutions of the brain; and the furrows, which separate them, and lodge the folds of the internal plate of the arachnoid, are called anfractuosities. They

A, anterior lobe of the brain, — *B,* median lobe, — *C,* posterior lobe, — *D,* cerebellum, — *E,* spinal marrow, — *f,* section of the corpus callosum, situated at the bottom of the fissure which separates the two hemispheres of the brain ; below this transverse band of white matter are found the lateral ventricles of the brain, — *g,* optic lobes concealed below the anterior face of the brain, — 1, olfactory nerves, — 2, the eye, in which terminates the optic nerve, the root of which may be followed upon the sides of the annular protuberance to the optic lobes; behind the eye we see the nerve of the third pair, — 4, nerve of the fourth pair, which is distributed like the preceding to the muscles of the eye, — 5, superior maxillary branch of the nerve of the fifth pair, — 5', ophthalmic branch of the same pair, — 5", inferior maxillary branch of the same nerve, — 6, nerve of the sixth pair going to the muscles of the eye, — 7, facial nerve, below the origin of this nerve we see a section of the acoustic, — 9, glosso-pharyngeal or nerve of the ninth pair, — 10, pneumogastric, — 11, hypoglossal, or nerve of the eleventh pair, — 12, spinal or nerve of the twelfth pair, — 14, and 15, cervical nerves.

vary in depth, and it is worthy of remark, that in the child and in most animals, even those nearest to man, the convolutions are but little developed. On the inferior face of the brain we find in each hemisphere three lobes, separated by transverse furrows, and designated as the anterior, median, and posterior lobes ; we also observe in this part of the brain two round eminences, placed near the median line (*mamillary eminences*), and two large peduncles which seem to issue from the substance of this organ to continue with the spinal marrow (commissures or penduncles of the brain). Also from this part of the brain issue the nerves, to which this viscus gives origin.

The surface of the brain is almost entirely formed of a gray nervous substance ; but its interior is composed of a white substance. When this organ is cut into, we find that there exist in its interior several cavities, which communicate externally, and which are called the ventricles of the brain. (Fig. 29, *f*).

Cerebellum. The cerebellum is placed beneath the posterior part of the brain (fig. 29, *D*, and fig. 28, *e*,) and has not one third the size of this organ, even in the adult, where it is larger in proportion than in the infant. It is divided like the brain, into two hemispheres, or lateral lobes, separated by a fissure, and a median lobe, situated behind and below, in a depression already mentioned. The surface of the hemispheres and of the median lobe is formed of a gray material, and presents no convolution, but a great number of furrows nearly straight, and placed parallel one by the side of the other, so as to divide this organ into a multitude of plates, arranged like the leaves of a book. Inferiorly the brain is continuous with the spinal marrow, by means of two short thick peduncles, and at the same point, it surrounds this latter organ with a band of white substance, which extends transversely from one hemisphere to the other, passing in front of the spinal marrow, with which it is closely united, and which bears the name of annular protuberance, or *pons Varolii*; so called in honor of a celebrated anatomist, Varolius.

When the posterior lobes of the brain are Optic lobes. raised, there may be seen between this organ and the cerebellum, four small round eminences, placed a pair on either side of the median line (fig. 29, *g*). They are raised upon the superior face of the medullary prolongations going from the brain to the spinal marrow, and constitute what anatomists call the optic lobes, or *tubercula quadrigemina*, to be noticed hereafter.

The *spinal marrow* (fig. 28, *f*, and 29, *E*,) is Spinal marrow. in some sort a prolongation of the cerebrum and cerebellum. It has the form of a thick cord, and presents in front, as well as behind, a median and longitudinal furrow, which divides it into two lateral and symmetrical halves. At its superior extremity (to which anatomists have given the name of *medulla oblongata*), several enlargements may be observed, called olivary, pyramidal, and restiform bodies, and from either side of which issue a great number of nerves, at first directed outward but afterwards descending more obliquely, so that the spinal marrow appears to terminate by dividing into a great number of longitudinal filaments, arranged as the hairs in the tail of a horse, (fig. 28, *j*), a resemblance, which has secured to this part the name of the object to which it has been compared. On a level with the origin of the nerves distributed to the thoracic members, the spinal

21

marrow presents an evident enlargement; afterwards it contracts, and again enlarges in volume at the origin of the nerves supplying the abdominal muscles; its lower extremity is very small, and is found toward the superior part of the lumbar region of the vertebral column.

The spinal marrow is composed, like the brain and cerebellum, of two medullary substances different in color; but here the gray matter, in place of being upon the surface, occupies the interior of the organ, and is covered by the white material.   There is no pia mater around the spinal marrow, and the sheath formed by the dura mater, is not entirely occupied by this organ, but is distended by a considerable quantity of liquid, in the   · midst of which it is suspended ; an arrangement admirably adapted to preserve it from the pressures or commotions, which might result from too violent movements of the vertebral column, or from any other cause, and which would produce upon this part of the nervous system consequences more serious than upon the brain.

Fibres of the encephal[o]n.   We have said, that the substance forming the cerebro-spinal axis was soft and pulpy; yet in the white matter fibres may be distinguished, and the study of their course leads to the explanation of many remarkable phenomena.

The spinal marrow presents, then, two halves, united by bands, formed principally of transverse medullary fibres ; on either side we find also, in the white substance of this organ, a great number of longitudinal fibres, which, at the superior part, are united in six principal portions.   Four of these portions occupy the anterior face of the medulla oblongata ; they constitute the enlargements known as the anterior pyramids, and olivary bodies, and they penetrate into the brain. The fibres of the pyramids present a very remarkable peculiarity; those of the right side branch to the left, and those on the left to the right : not till after this interlacing are they buried in the annular protuberance, and, continuing their course forward they constitute the peduncles of the brain. These fibres afterwards diverge and expand in the inferior, anterior and superior convolutions of the anterior and median lobes of the brain. The longitudinal fibres from the olivary eminences ascend, like those of the pyramidal, across the annular protuberance, and go to form the posterior and internal part of the peduncles of the brain ; they traverse, like those of the pyramidal bodies, various masses of the gray substance, increase in volume and number, and by following different directions form several parts of the brain, as the *thalami* of the optic nerves, and the *corpora striata ;* finally, they expand into the convolutions, the entire mass of which constitutes the cerebral hemispheres ; by the intervention of other transverse fibres the two halves of the brain communicate together, and these fibres constitute the corpus callosum already mentioned, as well as several other transverse bands, designated by anatomists under the general name of commissures.

The longitudinal fibres of the posterior pyramids of the spinal marrow are united to some fibres coming from the neighboring parts of the medulla oblongata, and thus constitute the peduncle of the cerebellum ; they plunge to the very centre of the corresponding hemisphere of this organ, and send to its circumference a multitude of plates which subdivide and form by their ramifications, as it were, branches enveloped in the gray material, and called by some anatomists the *arbor vitæ* (fig. 29, *D*). The communication of the two hemispheres of the cerebellum by means of transverse fibres may also be seen, a part of which surround the medulla oblongata in front and form the annular protuberance.

Nerves.   The *nerves*, which spring from the encephalon, and which establish the communication between this system and the different parts of the body, are in number forty-three pairs (fig. 28, and fig. 29). They all arise from the spinal marrow or base of the brain, and are distinguished, from their position, by regular numbers from before backward.   Most of them are at first formed by several roots, or collections of isolated fibres, and present near their origin an enlargement called a ganglion (fig.

Fig. 30.[1]   30, *c*). The twelve first pairs spring from

the encephalon, and issue from the skull by the various holes in the base.   The succeeding thirty-one pairs arise from that portion of the spinal marrow which is contained in the vertebral canal, and issue from this osseous covering by holes situated on each side, between the vertebræ. (Fig. 27.)

Finally, these nerves, with very few exceptions, soon divide into a multitude of branches, which in their turn are subdivided into twigs, and terminated by filaments extremely tenacious, in the substance of the different

[1] Section of the spinal marrow to show the disposition of the nerves originating from it, — *a*, spinal marrow, — *b*, the anterior root of one of the spinal nerves, — *c*, ganglion, situated in the course of this root, — *d*, posterior root of the same nerve, united to the anterior root beyond the ganglion, — *e*, common trunk formed by the union of these two roots, — *f*, small branch anastomosing with the great sympathetic nerve.

organs. Often, too, some of these nervous branches are united, and form an anastomosis[1] or a plexus.[2]

All the spinal nerves originate by two roots, composed of several fasciculi of fibres ; one from the anterior, the other from the posterior part of the spinal marrow (fig. 30).

The *ganglionic nervous system,* called also *the* Ganglionic system. *great sympathetic,* or *nervous system of the organic life,* is composed of a certain number of small nervous masses, perfectly distinct, but united to each other by medullary cords and various nerves, which anastomose with those of the cerebro-spinal system, or are distributed to the neighboring organs. These nervous centres are called ganglions ; they are found in the head, neck, thorax, and abdomen. Most of them are placed symmetrically upon either side of the median line in front of the vertebral column, and thus form a double chain from the head to the pelvis ; but they are also found in other parts, near the heart and in the neighborhood of the stomach.

The nerves of the cerebro-spinal system go to the organs of the senses, to the skin, muscles, &c. ; those which make a part of the ganglionic system are distributed to the lungs, heart, stomach, intestines, and sheaths of the blood vessels. In a word the former belong specially to the organs of relation, the latter to the organs of nutrition.

---

[1] The nerves, having been regarded by some anatomists as canals to conduct the nervous fluid, the name of anastomosis has been given to the union of their branches, or twigs ; this word, as we have already said, signifies really an emptying or a communication between two vessels.

[2] Plexus (from *plecto,* I interlace) is the name given to a kind of union of several nerves or vessels. The principal nervous plexi are those formed by the nerves of the limbs, soon after issuing from the vertebral column. Fig. 28, *g* and *h.*)

Such are the various parts which compose the nervous apparatus of man ; let us now see its uses, beginning with the study of sensation.

Sensation we have defined to be the faculty of receiving impressions, and being conscious of them. It belongs to all animals, but the degree in which it is developed varies with each of them. As we ascend in the scale of animal life, and approach man, the sensations are much more varied. The animal acquires the power of taking cognizance of a greater number of the properties possessed by surrounding objects, and of better appreciating slight shades of difference. The impressions produced become more lively, and as the faculty of sensation is thus perfected, the structure of the organs of the life of relation, become more complicated ; for here, as well as in all the other functions, by the division of labor does nature obtain increasingly perfect results.

Wherever the sensations produced by external objects are a little varied, there exists a distinct nervous system, and upon its action depends the faculty of perception. Its structure is at first very simple, and then all the parts composing it appear to fulfil nearly the same function. In the earth worm, for example, it is a knotty cord extended the length of the body, all the parts of which possess the same properties ; for if the animal be divided transversely into several sections, each of these fragments will continue to move and feel as before ; but in beings of a more complicated organization, and more perfect faculties, this apparatus is composed, as in man, of several dissimilar parts, each of which then acts in a manner different from the rest, and discharges special functions.

The most general phenomena, of those de- <small>Sense of tact.</small>
pendent upon the action of the nervous system, is the
perception of a sensation from the contact of a material
object with one of the organs of the animal. These are
not the sole sensations that may be experienced, and to
be more precise in our language, we must distinguish by
a particular name the faculty on which they depend, and
we therefore call it the sense of tact.

All parts of our body are not equally endowed <small>Sensible and Insensible parts.</small>
with this sense ; some organs enjoy a most ex-
quisitely delicate sense of tact, while others may touch
foreign bodies, and even be cut and torn by them, without the least sensation to the animal. Now, the most
sensible parts are most abundantly supplied with nerves,
and where there are no nerves there is no sensation. If
an incision be made in the paw of a living animal, and
the nerve of this part be exposed, it will be found that
this cord is endowed with an extreme sensibility ; however slightly it is pinched or pricked, the animal displays
all the signs of most acute pain, and the muscles, to
which the nerve thus injured is distributed, are agitated
by convulsive contractions.

From this it may be supposed, that to the <small>Influence of the nerves upon sensation.</small>
nerves our organs owe their sensibility ; and to
prove this it is only necessary to destroy one of these
cords. For if the experiment be performed upon the
limb of a living animal, all the parts supplied by the
nerve are immediately paralyzed, or deprived of the faculty of sensation and motion.

But is this nerve, whose action is indispensable to the

exercise of these functions, itself charged with determining the motions, and perceiving the sensations? Or
does it merely play the part of a conductor, and is it
destined only to transmit to the muscles the influence of
volition, and to convey to another organ, which may be
the seat of the perception of sensations, the impressions
resulting from the contact of an external object with the
surface of the body?

To solve this question physiologists have had recourse
to experiments upon living animals.

If, in any point of its extent, the nerve going to the
posterior claw of the frog, for example, be divided, and
the extremity thus separated from the rest of the nervous system, be pinched or pricked, it is found to be
completely insensible, while the part situated above the
section preserves all its sensibility; the parts of the
limb, which receive branches from the inferior fragment
of the nerve, are likewise paralyzed.

A nerve, separated from the system of which it made
a part, ceases then to discharge its functions; it cannot
consequently continue to be the seat of the perception of
sensation, and the conclusion must necessarily be drawn,
that it serves to transmit to the organ charged with this
function, the impressions received by the parts endowed
with the sense of tact.

This has been clearly demonstrated by all the researches made relative to this end. The impression
produced by the contact of a body with the nerve itself,
or with the part in which the nerve is ramified, cannot
be perceived, and cannot consequently produce a sensation, if it is not transmitted by the nerve to other organs.

This fact once established, we are naturally led to

ask, to what part must the sensations be conveyed to render the animal conscious of them, or, in other words, what is the organ charged with perceiving them?

The nerves, the functions of which we have now been studying, all end in the spinal marrow, and the latter terminates in the brain ; it is then evident, that this faculty must reside in some part of the encephalon. Let experiment decide for us, whether it is in the spinal marrow, cerebellum, or cerebrum. *Functions of the spinal marrow.*

When experiments are made upon the spinal marrow, similar to those already performed upon the nerves originating from it, the first observation is, that this organ is extremely sensible : the slightest prick produces an acute pain and convulsive motions ; and if cut across, a complete paralysis of all parts supplied by the nerves originating below the section is the result, while those, whose nerves arise from that portion of the spinal marrow yet in communication with the brain, continue to enjoy the faculty of sensation and motion.

By care to maintain artificial respiration, so as to prevent the animal thus experimented on from perishing with asphyxia, in consequence of paralysis of the respiratory muscles, it may be proved, that all parts of the spinal marrow and medulla oblongata lose the faculty of determining voluntary movements, and of giving birth to sensations, as soon as they are separated from the brain, and therefore the conclusion must be drawn, that the faculty of perceiving sensations, or of determining voluntary movements, does not reside in them.

But with the brain it is quite otherwise. If we expose the two hemispheres of this organ in a living animal (a pigeon for example), and irritate their *Part taken by the brain in the perception of sensations.*

22

surface with the point of a cutting instrument, we shall
be struck with their insensibility; we may cut and tear
the substance of the brain, and the animal will not betray
the slightest sign of pain, not even appearing to be aware
of the mutilation going on; but if, as in the experiment
of M. Flourens, this organ be removed, the animal falls
into a state of stupidity, from which it cannot be roused.
Its whole body becomes insensible, its senses no longer
act, and if it moves, it is because impelled by some outward power, and apparently the will is not concerned in
the determination of any of its movements.

From this experiment it may be learned, that the action of the brain is indispensable to the perception of
sensations and manifestation of the will; and that the
impressions, received by the nerves, must be conveyed
to this organ that the animal may be conscious of them.

Review.   In the function of sensation there is then a
very remarkable division of labor.   The parts, which by
their contact with foreign bodies are susceptible of giving
birth to sensations, cannot themselves perceive these
impressions; and on the other hand, the organ, which
has for its exclusive function the perception of these impressions, cannot itself directly receive them; it is insensible, and can be excited only by the impressions transmitted by the intervention of the nerves.

Three properties then may be distinguished in the
apparatus of the sense of tact; viz.: 1st, the faculty of
receiving by contact with a foreign body such an impression, as will give birth to a sensation; 2nd, the faculty
of transmitting these impressions from the point of their
production to the organ charged with perceiving them;
3d, that of giving to the animal the consciousness of their
existence, or of perception.

From the experiments of M. Flourens, and some other physiologists, it would appear, that in man and animals approaching him, such as the mammiferæ and birds, this latter faculty resides essentially in the hemispheres of the brain.

The faculty of transmitting to this organ the impressions produced by the contact of a foreign *Nerves of sensation and motion.* body belongs to the nerves of the cerebro-spinal system and to the spinal marrow, though all nerves do not possess this property. All those which originate from the spinal marrow, possess at the same time both the power of transmitting to the muscles the influence of the will, and of transmitting to the brain the impressions alluded to; consequently, they are nerves of motion and sensation. But the roots which fix them to the spinal marrow, do not present the same *Functions of the anterior and posterior roots of the spinal nerves.* properties. All these nerves, as we have seen, originate by two orders of filaments, one from the anterior, the other from the posterior portion of the spinal marrow, and the interesting experiments of M. Magendie have shown, that the latter serve for the transmission of sensations, and the former for the influence determining the voluntary movements.

If the posterior roots of a spinal nerve are divided, this cord is immediately deprived of the faculty of transmitting impressions; the part of the body to which it goes becomes insensible, but the movements remain under the influence of the will: while the section of the anterior roots, the posterior being untouched, causes the loss of motion without destroying the sensation.

By the division of the posterior roots of all the spinal nerves, the animal is not prevented from executing voluntary motions, but its whole body, (except the head, the nerves of which originate in the interior of the cranium) is rendered completely insensible. The posterior roots are then nerves of sensation, and the anterior of motion, and by their union the cord thus resulting enjoys at the same time the two faculties.

Cephalic nerves. Among the nerves from the cephalic portion of the encephalon, some possess the same property with the spinal ; these are the facial or nerves of the fifth pair ; the pneumogastric or tenth pair ; and the suboccipital or twelfth pair. All these nerves originate by roots, one of which presents a ganglionic enlargement and belongs to sensation, and the other has no ganglion and belongs to motion.

The other cephalic nerves are but little or not at all sensible, and serve either for motion, or for the transmission of certain peculiar impressions, produced by light, sounds, &c., and to which we shall have occasion to return.

Anterior and posterior columns of the spinal marrow. The different parts of the spinal marrow do not possess in an equal degree the faculty of transmitting sensations, or of exciting motions ; sensation is very delicate on the posterior face of this organ, and much more feeble on its anterior.

Great sympathetic nerve. Finally the ganglionary nervous system is but little, or not at all, sensible : these ganglions may be pinched or cut, as well as the nerves proceeding from them, without producing pain, or occasioning muscular contractions. It is also worthy of remark, that, in the state of health, the internal organs receiving these nerves transmit to us very feeble and confused sensations, and their sensibility is only developed in certain diseases.

In this case it is to be presumed, that the sensations arrive at the brain only by the intervention of the branches, which unite the nerves of the ganglionary system to each of the spinal nerves. But this point in physiology calls for new investigations.

Till now we have been occupied only by the Special senses. sensations produced by the contact of a material body upon our organs and with the function which permits us to recognise the existence of objects, which resist our movements, and to judge of the consistence, degree of polish, temperature, and, to a certain point, form of these objects. But these are not the only sensations these objects awaken in us, and we also enjoy the faculty of appreciating several of their qualities, which completely escape the sense of touch, such as their taste, odor, colors, and the sounds they produce.

These faculties constitute the special senses of man and most animals. For the perception of these sensations, as in the exercise of touch, the concurrence of the brain must be obtained to judge of them, and of a nerve to transmit to this organ the sensation produced ; but the division of labor is here carried yet farther, for this nerve is not fitted itself to receive the impression : the latter must be received by a special instrument and transmitted by the nerve of this organ to the brain. Thus the light, to produce upon us any sensation, must necessarily strike upon a determined part of the body, the sensibility of which is modified, although the nerve which conducts to the brain the impression thus received, is itself insensible to this agent. The nerves of the special senses all originate from the brain, or the neighboring part of the medulla oblongata, and enjoy but in a very limited degree

the sense of touch.   The optic nerve may be pinched
and cut without producing pain.   But the organs, which
are the seat of these special senses, and are all lodged in
the head, receive branches from the trifacial nerve, which
give to them their sensation.

These different modifications of the faculty of sensation
constitute the five senses, by means of which we acquire
all the ideas we have of surrounding objects.

Let us now examine in a particular manner how each
of these faculties is executed, and let us study the instruments, which serve both for touch, taste, smell, hearing,
and sight.

THE SENSE OF TOUCH.

All animals enjoy a sense of touch more or less delicate, which is executed by means of the membrane covering the surface of the body.   To study it we must first
examine the structure of the skin.

_{Structure of the skin.}   In man the exterior surface of the body and
the cavities in the interior which communicate externally,
such as the digestive canal, &c., are covered with a tegumentary membrane varying in thickness, and very different from the substance of the parts covered ; which is
continuous throughout, and really forms but one whole.
Its properties however, are not always the same, and it
is known by different names, as it is folded inward to
line the different cavities, or spread upon the exterior
surface of the body.   The internal portion of this general tegumentary membrane is called the *mucous membrane*, and the external *the skin*.

The skin is composed of three layers : the dermis or
chorion, the rete mucosum (mucous net work), and the
epidermis.

The *dermis* forms the deepest and thickest layer Dermis. of the skin. It is a white, supple membrane, but very elastic and resistent. It is composed of a great number of fibres and lamellæ interlaced in a very serrated manner. Its internal face is united to the neighboring parts by a layer, varying in thickness, of cellular tissue, and at some points giving attachment to the muscular fibres, which serve to move it. Lastly, upon its surface are a great number of small reddish prominences, very sensible, and arranged in pairs, forming in certain parts of the body, such as the palm of the hands and extremities of the fingers, regular series. These bodies are known as *the papillæ of the skin,* and the dermis of the skin of certain animals, when prepared by tanning, constitutes leather.

The *rete mucosum* is a plexus of vessels, soft Rete mucosum. in consistence, covering the dermis, and filled with a pulpy granulated material, to which the skin owes its color, and the tint of which, of course, varies in the different races, and even in different individuals. In negroes this coloring matter is black; in Europeans, it is white. When the rete mucosum forming it has been destroyed, it is not reproduced, and for this reason the cicatrices in the skin are white, even in negroes.

Finally, the *epidermis* is a kind of semi-trans- Epidermis. parent varnish, covering the surface of the skin, to which it is moulded; it is not a sensible part, nor even living, but a matter secreted by the skin, and only takes a certain degree of solidity when dry. Thus, in those parts of the body, which are withdrawn from the action of the air, it is soft and indistinct; and in animals living in the water it is not solidified, except when converted into stony matter, as in lobsters and most crustacea.

Pores. Upon the surface of the epidermis may be found a multitude of little openings, called *pores of the skin*. They correspond to the summit of the papillæ already Perspiration. mentioned, and open a passage to the perspiration, which is an acid liquid formed by secretion, and not to be confounded with the moisture continually exhaled upon the surface of the skin, and which constitutes insensible transpiration. These pores are extremely small, and do not traverse the dermis, and must be considered as the secretory ducts of the organs secreting the sweat.[1]

Hair. We also find upon the surface other and larger openings, some of which give passage to the hairs, the mode of the formation of which we shall hereafter investigate ; and from others, there distils a fatty matter secreted by follicles lodged in the thickness of this mem- Nails. brane ; lastly, at some points of the surface of the body, we find issuing from the substance of the skin horny plates called nails, in nature similar to the hairs.

Seat of the sensation. The sense of touch resides in the dermis, and seems to belong especially to the papillæ covering its surface. All parts of the skin are not equally sensible, which depends not merely upon the number of nerves distributed to them, but also upon the thickness of the epidermis covering them.

The principal use, then, of the epidermis is to oppose the evaporation of the liquids contained in the body, and to protect the skin, properly so called, from the immediate contact of foreign bodies, so as to moderate the im-

---

[1] The perspiration is an acid liquid, as the urine and the gastric juice. To be convinced of it apply upon the skin, moistened by this secretion, a piece of paper colored blue by tournesol; its color will be changed to red, as always results from the action of an acid.

pressions produced by this contact. We have already seen, that this solid covering is insensible, and as it is always interposed between the dermis and external objects, by contact with which upon this membrane sensation is caused, it will be easy to understand, that the thicker the epidermic layer, the more is the dermis withdrawn from the action of foreign bodies, and the more obtuse the impressions it must receive. Now, in some parts of the body, the heel for example, the epidermis is very thick; while in others, the extremity of the fingers, the lips, &c., it is extremely thin. Wherever too the skin is exposed to friction, its epidermis is thickened. Every one knows that the layer formed in the hands of blacksmiths and similar workmen becomes thick, hard, and wrinkled. Lastly, in some animals the epidermis is encrusted with calcareous matter, and becomes entirely inflexible; in this case, the surface of the body is rendered completely insensible.

The sense of tact, as it exists in all parts of _{Tact and touch.} the surface of our bodies, is sufficient to make us acquainted with the consistence, temperature and some other properties of the bodies in contact with it. This sense is then only exercised passively, and may therefore be designated as *tact;* but at other times the part possessed of this sensibility is actively impressed; muscular contractions directed by the will multiply and vary the points of contact with an external object, and then we give to this sense the name of *touch.*

Touch is then only tact perfected and made active; but it can be exercised only by organs arranged so as to be moulded in some sort upon the objects to be handled.

In man, the hand is the special organ of touch, _{Apparatus of touch.}

23

and its structure is very favorable to the exercise of this
sense.   Its epidermis is thin, smooth, and very supple,
its chorion is abundantly supplied with papillæ and nerves,
and reposes upon a thick layer of very elastic fatty cellular tissue.   Finally, the mobility and flexibility of the
fingers are extreme, and the length of these organs is considerable ; now these circumstances are the more advan-
tageous, in that they tend to increase the sensibility of
the part, and permit it to be applied to all bodies, whatever be the irregularity of their figure.   But another organic disposition, which no less contributes to the perfection of our touch, is the power possessed by man of
opposing the thumb to the other fingers, so as to be
able to enclose small objects between the parts of the
hand, which are precisely those in which the sensation is
the most exquisite.

In most animals the organs of touch are arranged in a
manner much less favorable.   In the mammiferæ, for example, this sense becomes obtuse as the fingers become
inflexible, and are enveloped in nails, by which they are
armed.   Sometimes, however, the place of the hands is
supplied by other organs of no less perfect structure, as
the trunk of the elephant ; and finally, some animals
employ the tongue as an instrument of touch, while others
are provided with particular appendages, answering the
same purposes, and which are called tentaculæ, palpæ, &c.

Use of
this sense.   Touch enables us to appreciate more or less exactly most of the physical properties of the bodies, on
which it is exercised ; their dimensions, form, temperature, consistence, polish, weight, movements, &c.   This
sense is so perfect, that several philosophers of antiquity,
as well as of modern times, have considered it as more

useful than sight or hearing, and as being even the source of our intelligence. These opinions are evidently exaggerated, for the touch has really no prerogative over the other senses; and in some monkeys, whose intelligence is incomparably less developed than that of man, the organs of touch are as perfect as in the human body.

SENSE OF TASTE.

The sense of taste, like that of touch, is excited by the contact of external objects upon certain surfaces of our body; but it acquaints us with properties, which escape the touch, to wit, the tastes of bodies.

All substances do not act upon the organ of <small>Taste of bodies.</small> taste. Some are very sapid, others but slightly so, and many are completely insipid. The cause of this difference is unknown, but it is found, that generally those bodies which cannot be dissolved in water, have no taste, while most of the soluble are more or less sapid. Their solution appears even to be a necessary condition of their action upon the organ of taste; for when this organ is completely dry, the sensation of taste is no longer imparted; and there are substances, which, being insoluble in water, are insipid in their ordinary state, but which acquire a strong taste when dissolved in some other liquid, such as spirits of wine.

The knowledge of the taste of bodies serves <small>Seat of the taste.</small> principally to direct animals in the choice of their food: and therefore the organ of taste is placed always at the entrance of the digestive tube. The tongue is the principal seat of it, but the other parts of the mouth may also experience the sensation of certain tastes.

The mucous membrane covering the tongue <small>Structure of the tongue.</small>

is abundantly supplied with blood vessels, and presents on its superior surface many eminences of various forms, which render it rough. These eminences, or papillæ, are of various kinds; some lenticular, and few in number, consist of a collection of mucous follicles; others, fungiform or conical and very numerous, are vascular or nervous; the last cover the terminations of the lingual nerve, and appear principally to serve for the sense of taste.

Nerves of the tongue. The tongue, the substance of which is formed by many muscles interlaced, receives the branches of several nerves; some serving to excite motion, others to conduct to the brain the sensation of the tastes. The trifacial nerve or the nerve of the fifth pair, which springs from the superior extremity of the spinal marrow, and separates from the encephalon near the anterior border of the annular protuberance, (see fig. 29) supplies these latter functions. It issues from the cranium behind the orbit, and divides into three principal branches, to wit; the opthalmic nerve supplying the apparatus of vision, etc., the superior maxillary nerve, which is distributed to the upper jaw and cheek, and the inferior maxillary nerve, one of the principal branches of which bears the name of the lingual nerve, and terminates in the mucous membrane of the tongue.

If the lingual nerve be divided in a living animal, the motions of the tongue are not paralyzed, but the organ is rendered insensible to tastes; and if the trunk of the trifacial nerve be divided in the interior of the cranium the sense of taste is destroyed, not merely in the tongue, but in all the other parts of the mouth.

The section of the nerves of the ninth and eleventh pair, which also go to the tongue, does not deprive the animal of the faculty of perceiving tastes, but occasions the loss of motion in the tongue and the other parts, to which these nerves are distributed.

It follows then that the lingual branch of the fifth pair is the special nerve of taste.

SENSE OF SMELL.

Certain bodies possess the peculiarity of exciting in us sensations of a particular nature, which cannot be perceived by the aid of the sense of touch or taste, and which depend upon the odor they exhale.

Odors are produced by particles of an extreme   Odors. tenuity, which escape from odoriferous bodies, and are diffused in the atmosphere, as vapors. All volatile or gaseous bodies are not odoriferous; but, in general, those which may be easily transformed into vapor, diffuse little or no odor, and in most cases we find, that odoriferous substances become more so when the circumstances in which they are placed favor volatilization. But yet the quantity of matter thus diffused in the air, to produce even the strongest odors, is extremely small. A particle of musk, for example, may perfume the air of an apartment for a long time, without any sensible change of weight. A multitude of bodies, such as water, clothing, etc., may imbibe these vapors, and in their turn become odoriferous; but other substances, such as glass, completely oppose their passage. We may perceive the odor of bodies at a great distance from us; but to arouse our olfactory sense, odoriferous particles, emanating from these bodies, must arrive in contact with the organ destined to receive them. And in this the mechanism of smell is analogous to the taste and touch, while with sight and hearing, as we shall soon see, it is quite otherwise.

Olfactory apparatus   The air is called the vehicle of odors; by this fluid they are transported to a distance, so as to reach us. It is then plain, that the organ destined to perceive them must always be placed so as to receive their contact, and experience teaches us, that to have this organ discharge its functions, the membrane touched by the odors must be continually moistened and covered by a liquid proper to absorb the odoriferous particles, and to fix them for a certain time upon its olfactory surface. If this surface were exterior, the former of these conditions would be fulfilled, but not the latter; the odors would strike it, but it would soon dry up, and become insensible to their contact. Smell must consequently reside in the walls of a cavity within the body, communicating freely externally; and the more rapidly and regularly the air, conveying to us the odors, is renewed, the more favorable are the conditions to the exercise of this sense.

This actually takes place, not merely in man, but also in all the other mammiferæ; in birds and reptiles also the sense of smell has its seat in the nasal fossæ, and these cavities are constantly traversed by the air, going to the lungs to supply the requisitions of the respiration. They communicate outwardly by the nostrils, and open posteriorly into the pharynx, at a short distance from the glottis. (See fig. 23.) Thus whenever the mouth is shut, the air must pass through them to reach the latter opening, and they may be considered as the anterior portion of the aërial tube.

The nasal fossæ are separated from each other by a vertical partition, directed from before backward, and occupying the median line of the face; their walls are formed by various bones of the face, and by the cartilages of the nose, and their

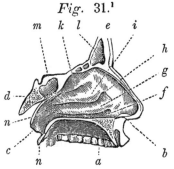

Fig. 31.[1]

extent is very considerable. Upon the external surface may be remarked three projecting plates curved upon themselves, and called the *ossa turbinata* (*g, i, k,*). They increase the surface of this wall, and are separated from one another by a longitudinal furrow, called a *meatus* (*f, h,*). Lastly, these fossæ communicate with *sinuses* hollowed out in the os frontis,[2] ossa maxillaria superiora, etc. The mucous membrane, lining the nasal fossæ, is called the *pituitary membrane;* it is thick, and prolonged beyond the borders of the ossa turbinata, so that the air can traverse the olfactory cavities only by narrow and long routes, and the least swelling of this membrane renders the passage of this fluid difficult, or even impossible. The surface of the pituitary membrane presents many little projections, which give it a velvety aspect;

---

[1] This vertical section of the nasal fossæ represents the exterior wall of one of these cavities; — *a*, mouth, — *b*, nostril, — *c*, posterior opening of the nasal fossæ, — *d*, portion of the base of the cranium, — *e*, forehead, — *f*, inferior meatus, — *g*, inferior turbinated bone, — *h*, median meatus, — *i*, median turbinated bone, — *k*, superior turbinated, — *l*, frontal sinus, — *m*, spheroidal sinus, — *n*, outlet of the eustachian tube.

[2] The *frontal sinuses* do not exist in infancy, but are developed by age and acquire very considerable dimension; these cavities cause the projection of the inferior part of the forehead above the root of the nose.

lastly it is continually lubricated by a liquid more or less viscous, called the *nasal mucus*; which appears to be formed in a great measure in the sinuses already mentioned, and it receives a very great number of nervous filaments, some from the fifth pair, and others from the olfactory or first pair.

Mechanism of the smell. The mechanism of smell is very simple ; it is only necessary, that the nasal mucus should imbibe the odoriferous particles, diffused in the air traversing the nasal fossæ, and that these particles should be thus arrested upon that part of the pituitary membrane which receives the filaments of the olfactory nerve. From this it can easily be conceived, how important is the nasal mucus to the sense of smell, and how the changes in the nature of this liquid, which take place during coryza or cold in the head, may cause, for a time, the loss of this sense.

In the superior part of the nasal fossæ the branches of the olfactory nerve are most numerous, the nasal mucus most abundant, and the routes followed by the air most contracted. In this part, therefore, are the odors most easily and vividly perceived. It would even appear, that the principal use of the nose is to direct the inspired air to the vault of the nasal fossæ ; persons losing this organ at the same time lose the sense of smell, and cases have occurred, where it was sufficient, to adjust an artificial nose upon the face of the patient, in order to restore this sense.

Nerves of smell. The olfactory nerve has generally been considered as the nerve destined to convey to the brain the impressions produced by odors ; but the nerve from the fifth pair appears to be of important service in the discharge of this function, for M. Magendie has proved that its section rendered the pituitary membrane insensible to the strongest odors.

With regard to the uses of the sinuses which Uses of the sinuses. communicate with the nasal fossæ by narrow openings, and which are lined by a thin membrane, nothing positive is known. It has been remarked, however, that animals, in whom these cavities are more vast, are also those in whom the sense of smell is the most delicate.

SENSE OF HEARING.

Hearing is the function which makes us acquainted with the sounds produced by vibrating bodies.

The apparatus of hearing is very complicated ; Auditory apparatus. the different parts of which it is composed are for the most part extremely small ; thus it occupies but a very small space, and is almost entirely contained in the interior of a bony prominence, which from either side of the head, advances into the interior of the cranium, and constitutes that part of the temporal bone called, from its hardness, the *petrous portion.* (Fig. 32, *e.*)

It may be divided into three portions, namely, the external, middle, and internal ear.

The external ear is composed of the pavilion External ear. of the ear, and auditory canal.

The pavilion of the ear (*a*) is a fibro-cartila- Pavilion of the ear. ginous plate, supple and elastic, which is perfectly free in the greater part of its extent, and which adheres to the edge of the auditory canal. The skin covering it is thin, dry, and very tense ; its surface turns in several ways, and presents various eminences and depressions, the most considerable of which is the concha (*d*). It

24

Fig. 32.[1]

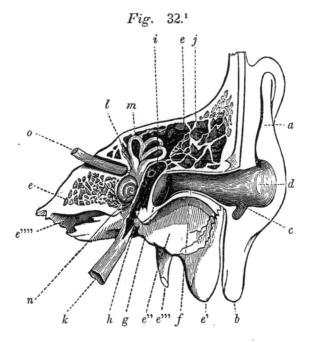

_{Auditory}
_{duct.} forms a sort of tunnel very open, and continuous
with the auditory duct, which is buried in the temporal

[1] This figure represents a vertical section of the auditory apparatus with
the interior parts slightly magnified, that they may be better seen. *a*, Pavilion of the ear, — *b*, lobule of the pavilion, — *c*, small eminence called the
antitragus, — *d*, concha, the bottom of which is continuous with the auditory duct, — *e, e*, portion of the temporal bone, called *the petrous*, in which
is lodged the auditory apparatus, — *e'*, mastoid portion of the temporal bone, — *e''*, portion of the glenroid fossa in which the lower jaw is
articulated, — *e'''*, styloid process of the temporal serving for the insertion
of the muscles and ligaments of the os hyoides, — *c''''*, extremity of the
canal traversed by the internal carotid artery before entering the cavity of
the cranium, — *f*, auditory duct, — *g*, tympanu n, — *h*, cavity from which
has been taken the chain of bones, — *i*, openii gs from the cavity of the
tympanum into the cells (*j*) of the petrous portion, — on the internal wall
of the cavity we find two openings, *fenestra ovalis, and rotunda*, — *k*, the
eustachian tube conducting from the cavity to the summit of the pharynx,
— *l*, vestibule, — *m*, semicircular canals, — *n*, cochlea, — *o*, acoustic nerve.

bone, and curves upward and forward. The skin lining this duct terminates abruptly at its internal extremity, and beneath it, we find many small sebaceous follicles, which furnish the yellow and bitter matter known as the *cerumen.*

The *middle ear* is composed of the cavity of the tympanum and the parts dependent upon it. Middle ear.

Fig. 33.[1]

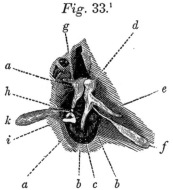

The *cavity* (fig. 32, *h*), is of an irregular form, hollowed in the petrous substance of the temporal bone, and making part of the auditory duct, from which it is separated by a membranous partition, very tense and elastic, called the *tympanum* (*b*). Opposite the opening, in which the tympanum is, as it were, set, (that is, upon the internal portion of the cavity), may be found two other holes, which are covered in the same manner by a tense membrane ; they are called from their form, *the oval* and *the round fenestra.* Upon the posterior wall of the cavity Cavity.

---

[1] This figure represents the external wall of the cavity, the tympanum, the bones of the ear, and their muscles, all enlarged. *a, a,* Frame of the tympanum, — *b,* tympanum, — *c,* handle of the mallet, its end resting upon the middle of the tympanum, — *d,* head of the mallet articulating with the anvil, — *e,* process, which springs from below the neck of the mallet, and buries itself in the glenoidal fissure of the temporal bone; its extremity gives attachment to the anterior muscle of the mallet, — *f,* internal muscle of the mallet, — *g,* anvil, a vertical branch of which rests upon the walls of the cavity, and a vertical branch articulates with the os orbiculare (*h*), — *i,* stirrup, the summit of which articulates with the os orbiculare, and the base of which rests upon the membrane of the fenestra ovalis ; — *k,* muscle of the stirrup.

is a hole conducting to the cells of the mastoid portion
of the temporal bone, and on its inferior wall may be seen
the opening of the eustachian tube, a long and narrow
duct, with an outlet to the posterior part of the nasal
fossæ, and which thus establishes a communication between the cavity of the tympanum and the external air.
Finally, this cavity is traversed by a chain of little bones,
which extends from the tympanum to the membrane of
the fenestra ovalis, and which leans, by means of a branch
directed to the side, upon the posterior wall of the cavity.
(Fig. 33.)

Fig. 34.[1]

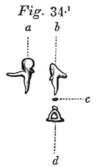

These bones are four in number, and
are called the mallet, or malleus (fig.
34, *a*), the anvil, or incus (*b*), os lenti-
culare (*c*), and stirrup, or stapes (*d*).
A small stalk, which may be compared
to the handle, and which belongs to the
malleus, rests against the tympanum, and
the base of the stapes thus reposes upon
the membrane of the fenestra ovalis.
Finally little muscles fixed to these bones, impress upon
them movements, by means of which they press more or
less strongly upon these membranes, and consequently
augment or diminish their degree of tension.

Internal ear.   The internal ear, as well as the middle, is entirely contained within the petrous portion. It is composed of several cavities, which communicate together,
and which are called the vestibule, semicircular canals,
and the cochlea.   The vestibule occupies the middle

---

[1] Bones of the ear separated, — *a*, malleus or mallet, — *b*, incus or anvil,
— *c*, os orbiculare, — *d*, stapes or stirrup.

part, and communicates with the cavity by the fenestra ovalis. The semicircular canals project upon the superior and posterior part of the vestibule; they are three in number, and have the form of rounded canals swelled at one extremity, like a flask. Lastly, the cochlea is a very singular organ, spirally twisted like the shell of the animal whence it takes its name. Its cavity is divided into two parts by a longitudinal partition, semi-osseous, semi-membranous; it communicates with the interior of the vestibule, and is separated from the cavity only by the membrane of the fenestra rotunda. This latter cavity is filled with air; the internal ear, on the contrary, is filled with an aqueous liquid, and the membrane lining the vestibule, as well as the semicircular canals, is not applied to the bony walls of the cavity, but is in a manner suspended in their interior.

The nerve of the eighth pair, which arises ^{Acoustic nerve.} from the medulla oblongata near the restiform body, and which departs from the encephalon between the peduncle of the cerebellum and the annular protuberance, enters the petrous portion through a bony passage called the internal auditory canal, and terminates in the interior of the membranous sacs of the vestibule, and semicircular canals, and also in the cochlea. Upon it depends the sensibility of the organ of hearing, and therefore it is called the acoustic nerve.

Such are the principal parts of the auditory ^{Mechanism of hearing.} apparatus. Let us now see the part taken by each in the exercise of the sense of hearing.

Hearing, we have said, is to make us acquainted with sounds.

Sound results from a very rapid vibratory mo- ^{Nature of sound.}

tion of the particles of sonorous bodies. To be assured
of this, it will suffice, merely to sprinkle some fine sand
upon a plate of glass, or the table of a violin, and to produce on this plate or instrument any sound whatever.
The grains of sand are at once agitated, and thrown into
the air with a force proportioned to the intensity of the
sound. The undulations experienced by the sonorous
body are communicated to the air in contact with its surface, as they were communicated to the grains of sand in
the preceding experiment; and thus from step to step
are sounds propagated to a distance. To perceive these
sounds, the vibrations now spoken of, must reach the
internal ear, and by their influence the liquid, which immediately bathes the acoustic nerve, will itself be thrown
into undulation. To point out the rationale of the mechanism of hearing, we must then follow the course of these
undulatory motions through the various parts of the auditory apparatus, interposed between the external air and
the acoustic nerve.

Use of the pavilion of the ear.   The sonorous vibrations of the air first strike
the pavilion of the ear. In animals, where this
part has the form of a horn, it serves to reflect the vibrations, and to augment the intensity of the sounds at its
contracted extremity, as experiment easily proves. Every
body knows that persons a little deaf hear with much
more facility when they use a similar horn, and if a thin
membrane be stretched over the open summit of a paper
cone, and its surface be sprinkled with fine sand, the
movements of the sand will be found much more intense,
when the sound arrives at the membrane by the broad
outlet of the tunnel, than when coming from the opposite
side.

In man, the concha and the auditory duct discharge the same functions ; but the other parts of the pavilion are not so arranged as to reflect sounds to the tympanum, and they also appear to have other uses. When sonorous vibrations fall perpendicularly upon an elastic surface, the undulatory motions, excited in the latter, are much more intense than when the sound arrives obliquely; and it may be concluded, that the varied directions of the surface of the pavilion of the ear, are destined to present to the sonorous waves, whatever be the direction in which they strike us, a plane thus arranged, and, consequently, they serve to augment the vibratory action of this elastic appendage. The pavilion of the ear is not, however, of very great utility, and the loss of it does not much affect the hearing.

The vibrations, thus excited in the pavilion of Use of the auditory duct. the ear, are communicated to the walls of the auditory duct, and thence to the deeper seated parts of the apparatus of hearing ; but these movements can only be very weak, and principally by the intervention of the air contained in this duct the sounds penetrate into the interior of the ear. Thus, if the tube be plugged with cotton or any other soft body, which opposes their passage, the perception of them is rendered indistinct.

· The tympanum serves principally to facilitate Uses of the tympanum. the transmission of sonorous vibrations from the external air to the acoustic nerve. The experiments of one of our most skilful physicians, M. Savart, prove that sounds, by striking upon a thin and moderately tense membrane, excite in it, very easily, vibrations. If a leaf of paper be stretched upon a frame, and its surface powdered with sand, the latter is readily agitated, and collected so as to

form varied lines, as soon as a sonorous body in vibration
is brought near.   If the same experiment be made with
a plane of wood, or a leaf of paper, no such movement
will result, unless the sound employed be extremely
intense.   But if to these latter bodies there be adapted
a membranous disc, similar to the tympanum, they will
readily vibrate under the influence of sounds, which before would have produced no appreciable effect.

It is plain, then, that the tympanum must readily vibrate, when sounds strike upon it, and that its presence
must augment the facility with which the other parts of
the auditory apparatus experience similar motions.

Transmis-
sion of sounds
through the ca-
vity. The vibrations are transmitted from the membrane of the tympanum to the bones of the ear,
to the walls of the cavity, and especially to the air, with
which this cavity is filled : they thus arrive at the posterior wall of the cavity, where there are membranes
stretched upon openings conducting to the internal ear,
nearly as the tympanum is stretched between the auditory duct and the cavity.   These membranes must act in
the same manner as the latter, that is, easily be made to
vibrate, and transmit these motions to the neighboring
parts.

Internal ear.   The posterior face of these membranous discs
is in contact with the aqueous liquid, which fills the internal ear, and in this liquid are suspended the membranous sacs,[1] which, in their turn, are distended by another
liquid, into which are plunged the terminal filaments of

---

[1] They are called the membrane of the vestibule, and the tubes of the
semicircular canals, according as they occupy the vestibule or the semicircular canals; in the cochlea there is nothing similar, and the liquid by which
it is filled, is the same which bathes the membrane of the vestibule.

the acoustic nerve. The vibrations executed by these membranes must then be transmitted to this liquid, afterwards be communicated to the membranous sac of the vestibule, and finally arrive at the nerve, upon which their action produces the sensation of sound.

From the preceding observations, it will be seen, that the air contained in the cavity plays Uses of the air contained in the cavity. a very important part in the mechanism of hearing; now, if this cavity did not communicate externally, this air would soon be absorbed and disappear, and the vibrations of the tympanum could be transmitted to the internal ear, only by the osseous walls of the cavity, and then with difficulty. This accounts for the use of the custachian tube, and explains to us in what way the obstruction of this duct may become a cause of deafness.

The tympanum is not indispensable to hearing, Utility of the cavity. for when this membrane is torn, the vibrations of the air contained in the auditory duct are communicated without interruption to the air of the cavity, and thus arrive at the membranes of the fenestra ovalis and rotunda. It might then be asked, what is the use of this, and what disadvantage could arise, if, the cavity not existing, the membranes of the fenestra ovalis and rotunda were placed externally? To reply to this question it must be borne in mind that the manner in which the membranes vibrate under the influence of the same sound, varies with their degree of dryness or humidity, their temperature, etc. Now it is probable, that two sounds make upon us the same impression whenever they cause the liquid, in which the acoustic nerve terminates, to vibrate in the same manner; and, consequently, in order that the same sound may always act upon us in an identical manner,

25

the membranes, which communicate directly their vibrations to this liquid, must constantly be at the same temperature, and the same degree of dryness ; and this is
precisely the case with the membranes of the fenestræ
of the internal ear.   The air of the cavity being renewed
but very gradually is always completely charged with
moisture, and at the same temperature, while if the cavity
did not exist, or had free external communication, the
condition of these membranes would be changed at every
instant, according as they were exposed to the action of
air, hot or cold, dry or moist.

Eustachian tube.   This also explains to us, why the eustachian
duct is long and narrow in all warm-blooded animals,
while in the cold-blooded, such as the lizards, it is short
and very large.   In the former, the air must have time to
ascend to the temperature of the body before penetrating
the cavity, while in the latter this temperature being the
same with the atmosphere, the speedy renewal of the air
contained in the cavity has no bad results.

Uses of the bones of the ear.   We learn, therefore, that the chain of bones
traversing the cavity, and which rests upon the
tympanum and the membrane of the fenestra ovalis, may
execute certain movements, by means of which the pressure it exercises upon these membranes may be increased
or diminished.   The utility of this arrangement is easily
understood ; if sand be sprinkled upon a membrane made
tense by a frame, and a sonorous body in vibration be
approximated to it, it will be found, that without in the
least changing the intensity of the sound, the violence
with which the sand is thrown into the air will be increased
or diminished, as the tension of the membrane is increased
or diminished.   In the former case, it will execute under

the influence of a sound of the same intensity vibratory movements much more extensive, than when the tension is increased. From this we may infer, that the pressure more or less strong of the malleus upon the tympanum, and of the stapes upon the membrane of the fenestra ovalis, will prevent these membranes from vibrating too strongly under the influence of very intense sounds, without depriving them of the faculty of vibrating, when a feeble sound strikes them. The pressure exercised upon the membrane of the fenestra ovalis is thus communicated to the membrane of the fenestra rotunda, by the intervention of the liquid with which the internal ear is filled ; and the result is, that the bones of the organ of hearing, by leaning upon the two membranes to which they are fixed, prevent the sonorous vibrations arriving at the acoustic nerve, from being so intense as to injure this delicate organ.

The loss of the malleus, incus, or os orbiculare diminishes the hearing, but does not destroy it; that of the stapes is, on the contrary, followed by deafness, for this bone, adhering to the membrane of the fenestra ovalis, by its fall tears the partition, and thus, the liquid contained in the vestibulum being lost, the acoustic nerve can no longer discharge its functions.

We see then, that all the parts composing the external and middle ear serve to perfect the hearing, without, however, being absolutely necessary to the exercise of this sense. Thus they gradually disappear, as we depart from man, and study the structure of the ear in animals gradually descending in the scale of beings. In birds there is no pavilion of the ear ; in reptiles the external auditory duct is also wanting, the tympanum becomes external and the structure of the cavity is simplified; finally, in most fishes there is neither concha, external, nor middle ear.

In animals placed yet lower in the series of beings, it is the same with the cochlea, and the semicircular canals, parts, the uses of which are not well known;[1] but the membranous vestibule is an organ never wanting; wherever there exists an auditory apparatus, there is always found a small membranous sac filled with liquid, in which the acoustic nerve terminates, and this vestibule is always an instrument indispensable to the exercise of the sense of hearing.

SIGHT.

Sight is that faculty, which makes us sensible of the action of light, and which acquaints us through this agent, with the form of bodies, their color, size and position.

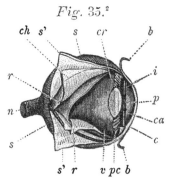

Fig. 35.[2]

Apparatus of vision. The apparatus charged with this function is composed of the nerve of the second pair, of the eye, and the several parts destined for the protection or motion of this organ.

Globe of the eye. The globe of the eye, with which we shall first be occupied, is a hollow sphere, a little swollen in front, and

[1] From the experiments of M. Flourens it would appear, that the destruction of the semicircular canals does not destroy hearing, but renders it confused and painful.

[2] Interior of the eye, — *c*, transparent cornea, — *s*, sclerotica, — *s'*, portion of the sclerotica, turned outward to display the membranes situated beneath,

filled with humors more or less fluid. Its exterior envelope is composed of two very distinct parts, one white,
opaque, and fibrous, called the *sclerotica* (*s*); the other
transparent and similar to a plate of horn, therefore called
the *cornea* (*c*). The latter occupies the front of the eye,
and is, as it were, set in a circular opening of the sclerotica. Its external surface is more rounded than that of
the latter membrane, and it resembles a watch-glass applied upon a sphere, and projecting beyond its surface.

At a short distance behind the cornea, in the Iris.
interior of the eye, is a membranous partition (*i*), which
is stretched transversely, and fixed to the anterior border
of the sclerotica, quite around the cornea. This kind of
diaphragm, which varies in color with the individual, is
called the *iris*, and presents in its middle a circular opening named the *pupil* (*p*). Muscular fibres may be detected in the tissue of this organ, directed like rays from
the edge of the pupil towards the circumference of the
iris, and other fibres of the same nature, which are circular and surround this opening like a ring. When the
former contract, the pupil dilates, by the action of the
latter, it is contracted.

The space comprised between the cornea and Anterior chamber.
the iris constitutes the anterior chamber of the eye (*ca*).
By the opening of the pupil it communicates with the
posterior chamber, a cavity situated behind the iris, and
which is filled, as well as the former chamber, by the

— *ch*, choroid, — *r*, retina, — *n*, optic nerve, *ca*, anterior chamber of the eye
placed between the cornea and iris, and filled with the aqueous humor, —
i, iris, — *p*, pupil, — *cr*, crystalline lens placed behind the pupil, — *pc*, ciliary
processes, — *v*, vitreous humor, — *b*, *b*, portion of the conjunctiva which,
after having covered the anterior part of the eye, is detached from it to line
the eyelids.

aqueous humor, a perfectly transparent liquid composed of water, holding in solution a little albumen and a small quantity of salts, such as are met with in all the secretions of the animal economy. This humor is supposed to be formed by a membrane behind the iris, which pre-

_{Ciliary processes.} sents a great number of radiated folds, called *ciliary processes* (*pc*).

_{Crystalline.} Almost directly behind the pupil is a transparent lens, called the *crystalline (cr)*. It is lodged in a diaphanous membranous sac (the capsule of the crystalline lens), and appears to be the product of a secretion from it ; for when taken from the eye of the living animal without destroying its capsule, a new lens is found to take the place of the old. It is also remarked, that this body is composed of a great number of concentric layers, constantly increasing in hardness from the circumference to the centre, which agrees with our remarks upon the mode of its formation. Its posterior face is also much more convex than its anterior.

_{Vitreous humor.} Behind the crystalline lens we find a large gelatinous diaphanous mass, resembling the white of an egg, and enveloped in a membrane of extreme tenuity, a great number of lammellæ from which extend inward, so as to form partitions or cells. This membrane is called the *hyaloid*, and the humor found in it the *vitreous humor (v)*.

_{Retina.} Every where, except in front, where are the crystalline lens and the iris, the vitreous humor is surrounded by a soft white membrane called the *retina (r)*, which is only separated from the sclerotica by another

_{Choroid.} membrane, equally thin, which is called the *choroid (ch)*. The latter is principally formed by a net-work of blood vessels, and is stained with a black matter, which

gives to the bottom of the eye that deep color, which is seen through the pupil, and which is wanting in those persons and animals called *albinos.*

The globe of the eye receives several nerves. Optic nerve. The most remarkable for its size and functions is the optic nerve (*n*), which traverses the posterior part of the sclerotica, and is continuous with the retina, which appears in fact but an expansion of it. The other nerves of the globe of the eye are extremely small, and Ciliary nerves. are called the ciliary nerves : they spring from a small ganglion, formed by the union of some branches of the nerves of the third and fifth pairs (see fig. 29), and are distributed to the iris and the proximate parts of the interior of the globe of the eye.

By the intervention of light, we have said, Mechanism of vision. bodies placed around us act upon our sight. Those which emit light, the sun and bodies in ignition, for example, are visible of themselves ; but others become so, only when the light striking upon them is reflected in such a way as to meet our eyes.

This agent moves with an extreme rapidity ; it can act upon our senses only to the extent it strikes upon the retina, situated at the bottom of the eye. Opaque bodies reflect or absorb it, but transparent bodies, such as the atmosphere and water, afford it a free passage.

The first condition, then, for the exercise of vision is the absence of every opaque body between external objects and the bottom of the eye. Therefore the cornea, which covers the anterior part of this organ, like a watch glass, is completely transparent, and the light, which traverses it, and passes through the opening of the pupil, readily arrives at the retina ; for it only encounters on

the way the crystalline lens, which is diaphanous, and
the humors, which are equally so.

But in some diseases it is quite different, and this loss
of transparency always causes blindness. In the affec-
tion known -as the cataract, for instance, the crystalline
lens becomes opaque, and thus opposes the passage of
light: and when white spots or pellicles are formed upon
the cornea, this membrane becomes a kind of screen,
which prevents the luminous rays from penetrating to the
eye, and thus entirely destroys vision.

The diaphanous parts of the globe of the eye do not
serve to give passage to the light merely. Their principal use is to change the direction of the rays which enter
this organ, so as to collect them upon some point of the
retina. The interior of the eye resembles exactly the
optical instrument known as the camera obscura, and the
image of the objects seen by us is painted upon the retina, as upon the curtain placed behind the latter. To
understand this phenomenon, it is necessary to examine
the course of luminous rays through transparent bodies
in general, and to apply the knowledge thus acquired, to
the study of the mechanism of vision.

Light ordinarily advances in a straight line, and the
different rays, which start from any one point, are dispersed according to the direction in which they travel,
and the distance of the space traversed. When these
rays fall perpendicularly upon the surface of a transparent body, they traverse it without any change in direction ; but if they strike it obliquely, there is always more
or less deviation from their primitive course. If the body,
into which they penetrate, be more dense than that from
which they issue, if they pass from air into water or

glass, for example, they then form an elbow and approach the perpendicular at the point of immersion. If, on the contrary, they pass from a dense to a rare medium, they depart from this perpendicular, and these deviations are greater in proportion to the obliquity, with which the ray strikes the surface of the transparent body.

This phenomenon, which is known as the refraction of light, is easily understood. It is owing to this change in the direction of the luminous rays, in their passage from water into air, that a straight stick, plunged half its length into the former liquid, always appears as if bent at the point of immersion; and if a piece of money (*a*) be placed at the bottom of an empty vase, and the edge of the latter be raised just high enough to prevent the eye of the observer from perceiving this object, to render

Fig. 36.[1]

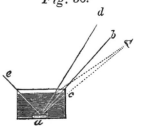

it visible, it will be sufficient, to fill the vase with water. For the ray of light coming from the money, instead of always advancing in a straight line, will be refracted in its passage from the water into air, and will depart from the perpendicular; and by this change in direction, the rays, which before passed above the eye of the observer, will strike upon it.

[1] From the position of the eye it is evident that if the light travelled in a straight line, the observer could perceive the piece of money (*a*) only so long as the ray *a, c,* reached his eye; but the walls of the base being opaque this ray and all others situated below the line *a, b,* and *a, c,* are intercepted. Now when the vase is filled with water the rays are refracted in passing from this liquid into the air, and consequently, one of the rays which before passed above the eye, the ray *a, d,* for example, will be deviated so as to reach the observer.

The luminous rays, we have observed, approach the perpendicular at the point of contact, whenever they penetrate obliquely a body denser than that from which they issue.   Therefore, the form of bodies exerts a great influence upon the course of the light traversing them ; and as their surface is convex or concave, these rays will be collected or dispersed.

Some examples will render this proposition easy to be understood.   Let us suppose three diverging rays to set out from the point *a*, traverse the air, and fall upon a lens with a convex surface, represented by the line *b*.   The

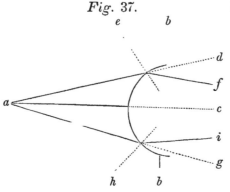

Fig. 37.

ray *a c*, will strike perpendicularly upon this surface, and consequently will traverse the lens without experiencing any deviation ; but the ray *a d*, falling obliquely upon this surface, will be refracted, and approach the perpendicular drawn to the point of immersion ; now this perpendicular will take the direction of the dotted line *e*, by approximation to which, the luminous ray, instead of following its course towards the point *d*, will follow the line *f*.   The same will be the case with the ray *a g*, which in continuing its course will approach the perpendicular *h*, and be directed towards the point *i*, instead of continuing in a straight line towards the point *g*.   The other rays, which strike upon the lens, will be refracted in an analogous manner, and, consequently, in the place of being dispersed, they will be collected and may even unite in one point, which may be called the *focus* of the lens.

If the surface of the crystal, in place of being convex, is concave, the luminous rays will not approach the axis of the bundle, as in the preced- ing case, but, on the contrary, will greatly diverge. The ray *a d*, for example, must approach the per- pendicular at the pendicular at the

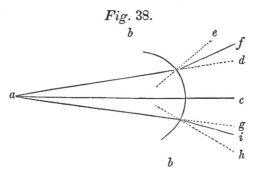

Fig. 38.

point of contact, which will have the direction of the dotted line *e*, and by this deviation the ray will take the direction of the line *f*.

The refraction of the luminous rays, in thus traversing convex or concave lenses, is greater in proportion to the curvature of the surface of these bodies, and the mere inspection of the figures we have made use of will be sufficient to make us comprehend it must be so. For the greater the curvature of the surface upon which the di- verging rays strike, the more will the perpendiculars, at the point of immersion, depart from these same rays.

Physics also teach us that transparent bodies refract the light more powerfully in proportion to their density, (that is to say when under the same volume they have a more considerable weight), and when formed of more combustible materials.

The light which strikes a transparent body does not entirely traverse it; a considerable portion is reflected, and owing to this property, bodies answer more or less perfectly the purpose of mirrors.

Course of the
light in the eye. · From what precedes, we see that when a collection of luminous rays falls upon the cornea, one part must be reflected while the other traverses it. It is the light thus reflected, which gives to the eyes their brilliancy, and which causes our own image in the eyes of others. The rays, which penetrate into this transparent plate, pass into a body much denser than the air; they are, consequently, refracted and approximated to the perpendicular or axis of the collection, the more forcibly in proportion to the convexity of the cornea; for the greater the projection of this membrane, the more acute will be the angle formed by the diverging rays striking upon its surface.

Uses of the
aqueous humor. If, after having traversed the cornea, the luminous rays entered the air, they would be refracted to the same degree as upon their entrance into this membrane, but in a contrary direction; they would consequently retake their primitive direction. But the aqueous humor which fills the anterior chamber of the eye, possesses a refracting power much greater than the air, so that on entering it the rays are less dispersed, than they were approximated by their passage through the cornea; the action of these parts therefore renders them less divergent than before their entry into the eye, and causes a more considerable quantity of light to pass into the opening of the pupil.

Uses of the
pupil. A great part of the light, which arrives at the bottom of the anterior chamber of the eye, meets the iris and is absorbed or reflected outwardly by it: that only, which falls upon the pupil, penetrates to the bottom of the eye, and the quantity is in proportion to the size of the opening. Thus when but a small quantity of light reaches the eye, the pupil is dilated, while it contracts under the influence of a brilliant light ; the iris, as we see, regulates the quantity of light reaching the retina.

The rays of light having traversed the pupil, _{Uses of the} _{crystalline.} fall upon the crystalline lens, which is diaphanous, and which changes anew their direction, causing them all to converge to one point, called the focus, where they unite. Now this focus is precisely upon the surface of the retina, and it is thus that the luminous rays, sent to the eye from different points of a body placed at a distance, are collected upon this nervous membrane so as to paint upon it, in miniature, the object from which they emanate.

It is easy to be convinced by experiment that _{Formation} _{of images upon} images are thus formed in the bottom of the _{the retina.} eye. Take the eye of a hare or pigeon, the sclerotica of which is nearly transparent, or better yet, the eye of an albino, and place in front of the cornea a very brilliant object, a lighted candle, for example, and the image of the latter will be seen depicted upon the retina.

These images are al-
ways formed upside down,
and the cause of this phenomenon is easily shown.

Fig. 39.

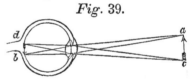

When we consider the course the luminous rays, origi-
nating from the two extremities (*a, c,*) of an object, must take to reach the retina, it will be seen they must cross before reaching it, and that consequently the ray coming from the superior extremity (*a*), of an object will be at the lower part of the space occupied upon the retina by the entire collection of rays forming the image (*b*), while that coming from the inferior extremity of the object (*c*) will occupy the top of the same space (*d*). The same

will take place with all the other rays, and therefore the object will appear reversed at the bottom of the eye.

Uses of the choroid. The black matter, which is situated behind the retina, and which lines the whole bottom of the eye as well as the posterior face of the iris, serves to absorb the light immediately after it has traversed the retina. If this light were reflected in other points of the membrane, it would considerably trouble the sight, and prevent the clear formation of images on the bottom of the eye. Thus in albinos, whether man or animal, where this pigment is wanting, vision is extremely imperfect; during the day they can scarcely see at all.

Perfection of the eye. The globe of the eye serves to conduct the light and to concentrate it upon the retina. It discharges the office of a kind of spectacle-glass, but it is an optical instrument far more perfect than any of those ever yet constructed by scientific men : for at the same time that it is perfectly achromatic and presents no aberration of sphericity, its capacity may vary considerably.

Achromatism. By *achromatism* is meant the property of turning light from its course without developing any color, and, consequently, the achromatic lenses are those which form in their foci colorless images, or possessing only the colors of the object represented. The white light results from the union of the seven colored rays of the solar spectrum, and these different rays are not equally refrangible. Wherefore, when light is passed through a refracting body, it is more or less completely decomposed, and the objects whence it proceeds, appear to have the color of the solar spectrum. Achromatic telescopes are obtained by combining several glasses, some of which correct the dispersal of the light produced by the others,

so as to unite in one focus all the rays. It is probable
the achromatism of the eye depends upon a similar arrangement, but physicians have not yet agreed upon the
explanation of this phenomenon; some think it depends
upon the different humors of the eye, others attribute it
to the difference of density in the different layers of the
crystalline lens.

The *aberration of sphericity* consists in the Aberration of sphericity.
union of rays which fall upon different parts of a lens,
into foci sensibly different, whence results a want of
clearness in the images. When the lenses are very convex, the rays which pass near the edges do not unite in
the same focus with those which traverse the central
part of the instrument, and to obtain clear images, the
passage of the former must be intercepted by placing in
front of the lens a separating medium pierced by a hole.
Now, the images formed behind the crystalline lens of
the eye are never diffused, and this want of aberration of
sphericity has been attributed to the iris, which answers
the purpose of the divisions in the interior of telescopes.

Every one knows that objects may be seen with the
same clearness when placed a few inches from the eye,
as when at a considerable distance from this organ. In
our optical instruments, on the contrary, the image
formed in the focus of a lens advances or recedes, according to the distance of the object. It has, therefore,
been supposed that to give to our sight such varying
capacity, the crystalline lens must approach or recede
from the retina as necessity required, or else the globe of
the eye must change its form. But direct observation
does not confirm these hypotheses, and this peculiarity
has never found a satisfactory explanation.

However, the eye does not always possess in the same degree this precious faculty; some can see distinctly only at the distance of some feet; nearer all images are confused. With others, on the contrary, the sight becomes clear only when the objects are brought within a few inches of the eye, and every thing beyond seems enveloped in a cloud.

Presbytia.   The former of these infirmities, known as *presbytia*, depends upon a want of convergence in the collections of rays traversing the humors of the eye. The rays which arrive at this organ from a very distant object, diverge very little and may be collected at one point of the retina, although the refracting power of the eye be small; but those which come from a near object diverge greatly, and the refracting power of the eye is too weak to collect them upon a determinate point of the retina. Those thus afflicted have usually a contracted pupil, as if they were continually making an effort to prevent any other rays from entering the eye than those falling upon the centre of the crystalline lens, and which do not require great deviation from their course to be collected together behind it upon a fixed point of the retina. This want of refracting power in the eye appears, in general, to belong to a flattening of the cornea or crystalline lens, which circumstance must tend to produce presbytia, and which may be found in nearly all old men.

Myopia.   *Myopia* is the result of a contrary effect. The rays which traverse the eye are then so forcibly deviated from their course, that in place of diverging they even converge before reaching the retina. This imperfection of the visual organ depends, in general, upon too great a convexity of the cornea or crystalline lens.

It is remarked that short-sighted persons become less so by age, which happens in consequence of a diminution in·the secretion of the humors of the eye always occurring in old age, which, by rendering the cornea less convex, renders the sight longer; in most cases, it causes presbytia, but here it only serves to correct the errors of the eye and to give to the sight its usual character. Therefore the vision of short-sighted persons is improved by age, while in others it is usually weakened; but as this diminution in the abundance of the humors of the eye always continues, there is a moment when the eye of the short-sighted individual becomes also too much diminished in the power of refraction, and his sight, consequently too long.

To correct these natural faults of the eye, recourse must be had to means, the efficacy of which confirms the explanation just given of the cause, whether of myopia or presbytia. Glasses are therefore placed before the eyes with surfaces so directed as to augment or diminish the divergence of the rays traversing them. Short-sighted persons make use of concave glasses, which give divergence to the rays of light, and the far-sighted employ convex, which, on the contrary, collect the rays diverging from the axis of the bundle.

The contact of light with the retina, we have ^{Uses of the retina.} said, causes vision, and when this membrane is paralyzed (a condition which constitutes the disease known as gutta serena), this sense is completely destroyed. But the sensibility of the retina is entirely limited, and can only be excited by this subtile agent. This nervous membrane enjoys little or no sensibility to the touch, and it

27

may be touched, pinched, or even torn upon the living animal without the slightest manifestation of pain.

However, this peculiar sensibility of the retina has its limits: too feeble a light does not act upon it, and too strong a light injures it and renders it incapable of action. But, in this respect, the influence of habit is extreme; when a person has remained sometime in obscurity, a light, although very feeble, dazzles the eyes, and renders, for some instants, the retina incapable of discharging 'its functions, while those accustomed to the light of day experience the same effects only when looking upon the most brilliant objects, in seeking, for example, to take the sun's altitude.

When we look for a long time upon the same object without a change of position, the point of the retina receiving the image is soon` fatigued, and this fatigue, carried beyond a certain limit, deprives for some time, the part, experiencing it, of its usual sensibility. Thus, if we look for some time at a white spot on a black ground, and then change our sight to a white ground, we think we see a black spot, because the point of the retina already fatigued by the white light has become insensible to it.[1]

The fatigue experienced by the retina in the exercise of its functions depends in part upon the efforts made to fix the attention upon the object placed before the eyes. If we endeavor to examine attentively bodies in a feeble light, we soon experience a painful sensation in the orbit, and also in the head.

[1] The black color depends upon the absence of light, and the bodies which afford it to us are those which absorb all the light falling upon them; we perceive them only because they are surrounded by bodies which reflect them.

All points of the retina are adapted to receive the impression of light; but the central part of this membrane enjoys a more exquisite sensibility than the rest, and it is only when the images of external objects are formed upon this part, that we see them distinctly. Thus, when looking at any object, we take care to direct toward it the axis of our eyes.

We might attain the same end by means of the various movements of which the head is susceptible; but, in order to render these changes in the direction of the eyes more easy, nature has provided these organs with muscles destined especially for their motion.

These muscles are fixed to the sclerotica by their anterior extremity, and inserted by their opposite behind the globe of the eye (to the bottom of the orbit), and, since this organ reposes upon fatty cellular tissue, without intimately adhering to it, each of these muscles by contracting, turns the eye to its own side. They are six in number: four of them called recti muscles, inserted

Muscles of the eye.

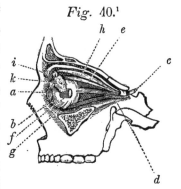

Fig. 40.[1]

---

[1] Vertical section of the orbit to show the position of the eye and its muscles. — *a*, Cornea, — *b*, sclerotica, — *c*, optic nerve, the opposite extremity of which enters the globe of the eye, — *d*, rectus inferior muscle of the eye, — *e*, rectus superior muscle, — *f*, portion of the rectus externus muscle of the eye; at the bottom of the orbit we see the other extremity of this muscle, of which the entire middle part has been removed to display the optic nerve situated behind it; — *g*, extremity of the small oblique muscle, — *h*, large oblique, the tendon of which passes into a small pulley before it is fixed to the sclerotica, — *i*, elevator muscle of the superior eyelid, — *k*, lachrymal gland.

in the four opposite points of the circumference of the sclerotica and proceed directly backward, so that they can direct the eye upwards, downwards, to the right or left, depending upon the muscles called into action. Lastly, two of these muscles which are called the oblique muscles, (*h, g,*) are so arranged as to cause this organ to execute movements of rotation, which direct the pupil downwards and inwards, or upwards and outwards.

Uses of the nerves of the eye. The nerves, which give motion to these muscles, belong exclusively to the apparatus of vision. They are furnished by the third, fourth and sixth pairs. (Fig. 29). The recti muscles are entirely under the direction of the will; the oblique often act independently of it, and from their contraction arises the rolling up of the eyes in syncope.

The optic nerve, which, by its expansion on the bottom of the eye, forms the retina, transmits to the brain the impressions produced upon this membrane by the contact of light: its section therefore produces total blindness.

In order that the retina may discharge its functions requires not merely the consent of the optic, but also of the nerves of the fifth pair, which we have already seen exercises the greatest influence upon taste and smell. If the section of this nerve be made between the brain and the point of origin of the branches to the eye, vision is destroyed. The animal appears to distinguish light from darkness, but is really blind ; and singular though it be, at the end of some time the cornea becomes opaque, ulcerates, the eye is emptied and atrophied.

Perception of images. The hemispheres of the brain appear to be the seat of the perception of these sensations, as of all others ;

for when they are destroyed the animal becomes at once blind. But there are other parts of the encephalon which also exercise great influence upon this sense; these are the optic lobes, or tubercula quadrigemina (fig. 29, *g*,). When destroyed in a bird (where these parts are much developed), blindness will also ensue; and it is worthy of observation that animals in whom the retina is most developed and the optic nerves the largest, are also those in which these lobes acquire the most volume and have the most complicated structure. These organs may even be considered as depending upon the optic nerve, and as being the uniting bonds of these nerves to the cerebral hemispheres.

But the most striking result of our experiments upon the encephalon is to find, that the destruction of the cerebral hemisphere or optic lobe of one side does not occasion the loss of sight of the same side: the eye of the opposite side becomes blind, and anatomy explains this fact; for the optic nerves soon after their separation from the brain, unite and interlace so that the one furnished by the right lobe goes to the left eye, and vice versa.

We judge of the form of bodies by that of the image they produce upon our retina; therefore, when, from any cause, the form of the luminous pencil of rays sent by them to this membrane, is changed before reaching the eye, we fall into great errors in our judgment. The experiment already cited of a stick half plunged into water and then appearing bent, although actually straight, is an optical illusion of this kind.

We judge of the position of surrounding objects by the direction of the luminous rays they send to

us, and we always see them in the extension of the
straight line followed by these rays at the moment they
penetrate the eye.   For this reason, when the pencil of
rays sent by one of these objects upon a polished surface
(a mirror, for example,) is reflected by the latter so as to
make a greater or less angle and arrive at the eye, we
see the object as if it were placed behind the mirror, in
the prolongation of the straight line pursued by the ray
meeting the eye from this instrument.   Judgment may
rectify the consequences we draw from this sensation;
but it always exists.

This accounts for our using but one eye when we wish
to be certain that bodies are in an exact line with each
other.   When this condition is actually fulfilled, and one
of our eyes placed upon the prolongation of the line
occupied by these objects, the luminous ray which is
directed from the more distant body towards our eye
cannot arrive to it, being intercepted by the intervening
body, and so of the rest, consequently the nearest body
conceals from us entirely or in part all the others; when
if regarded by both eyes, the same thing happens only
when the objects are so far distant from us that the rays
sent to our eyes are almost parallel, or when the intervening object is very large in comparison with the more
distant or very near to it, and even then the coincidence
is much less plain than if the observer uses but one of
his eyes.

Distance of
objects.   To estimate the distance, which separates us
from objects, the simultaneous action of the two eyes is,
on the contrary, of great assistance : the following experiment will convince us of the fact.   Suspend a ring by
a thread, and try to introduce a pointed instrument fixed

to the end of a long stick; by making use of both eyes
you will easily succeed at every trial; but if one eye be
closed you will find the greatest difficulty in threading
the ring: the instrument will go beyond or stop short,
and only by chance or after repeated trials will you succeed in introducing it into the ring.

Thus when an individual has lost one eye he generally
remains some time unable to judge correctly of the distance of bodies situated near him, and this privation renders the appreciation always much more difficult.

But the utility of the two eyes in this case is easily
explained upon physical laws. When an object is only
a short distance from us, it is requisite, in order that its
image may fall upon the same point of the retina of the
two eyes, that the axis of these organs should converge
toward the point looked at, and this inclination, of which
we are conscious, is great according to the proximity of
the object. But when objects are so far distant that in
beholding them the optic axes of the two eyes become
sensibly parallel, we have no longer a sure rule to determine their distance, and we can only form our judgment
from data more or less deceptive, such as the degree of
light, the clearness with which we distinguish the minu-
tiæ, the size of the object itself, our previous acquaintance with it, etc. When we can compare this distant
object with those intervening, this appreciation becomes
much more certain; every one knows how difficult it is
to judge of the distance of a light seen in the middle of
the night, when the obscurity prevents surrounding objects from being seen.

The concurrence of the two eyes is moreover essential
in that it makes objects appear plainer. If we regard a

strip of white paper with one eye, and place before the other an obstacle which will conceal half the object; the portion seen by both eyes, at the same time, will appear much more brilliant than when seen by one alone.

Size of objects. The manner in which we judge of the size of bodies depends much more upon intelligence and custom, than upon the action even of the apparatus of vision. Our first guide is really the size of the image formed at the bottom of the eye ; but as the distance between us and an object increases, this image diminishes, so that to judge of the dimensions of the former, the distance at which we suppose it must always be considered. When, therefore, the distance is not exactly appreciated, it is difficult to judge of the size of a body seen for the first time. A mountain, which we see afar off for the first time, appears to us, in general, much smaller than it really is, because we imagine it near us, when it actually is at a considerable distance.

Motion of objects. The estimation of the motion of bodies is sometimes made from the change of direction of the light which arrives at the eye, whence results the displacement of the image upon the retina, sometimes from the variation in size of this same image. In order to follow the motion of a body its displacement must not be too rapid, for then we do not perceive it unless it project a very great quantity of light, and, in this case, the same effect is produced upon our eyes as if it instantaneously occupied the whole length of the line traversed. On the contrary, we generally recognise with great difficulty, and sometimes we even cannot recognise the motion of bodies the image of which is very slowly displaced, either from the actual slowness of their motion, as in the hand of a watch, or from their great distance, as in the stars.

From the remarks made with regard to the manner in which we judge of the distance and size of bodies, it is easy to see that the sense of vision has need of a kind of education, and that under some circumstances it will always lead us into error.

It is by making use of these errors, known in physics and physiology by the name of *optical illusions*, as well as of the laws of the animal economy on which they depend, that the arts are able to produce them at will, to make plane surfaces appear round and prominent, and to make near objects more or less distant.

That sight may give us the valuable information it is susceptible of communicating, this sense must undergo long exercise and a thorough education. The new-born infant distinguishes only the light from the darkness; and although its eye already presents all the physical qualities necessary to vision,[1] it does not begin to see till after some weeks of existence. At first its eyes are only directed to the most brilliant objects, such as the sun, etc., without distinction. The first things which attract it are those of a red color; it soon, however, begins to distinguish the other brilliant colors, but has, as yet, no idea of distance or size, and it will extend its hand to seize upon remote objects, without the least regard to their dimensions. By degrees vision is perfected, and it is principally in correcting, by the aid of the other senses, the errors to which the former is exposed that the infant acquires the faculty of judging rightly of what he sees around him.

---

[1] In the fœtus prior to the seventh month, it is otherwise; the iris has not yet been perforated; but at this period the pupillary membrane, which occupies the place of the pupil, breaks and is absorbed so as to give access to the light.

To give some idea of the kind of education necessary
to vision, it will be sufficient to read the curious history
of a man blind from his birth, to whom Cheselden, a
celebrated English surgeon, restored sight at a sufficiently
mature age to enable the young man to analyze all his
sensations, and give an account of them.

" When he first saw, he was so far from making any
judgment about distances, that he thought all objects
whatever touched his eyes (as he expressed it) as what
he felt did his skin, and thought no object so agreeable
as those which were smooth and regular, though he could
form no judgment of their shape or guess what it was in
any object that was pleasing to him.   He had during his
blindness such feeble ideas of colors as he could distinguish by a strong light, but they did not leave sufficient
traces to be recognised ; when he did see them he said
the colors he perceived were not the same he formerly
saw.   He knew not the shape of any thing, nor any one
thing from another, however different in shape or magni-
tude : but upon being told what things were, whose form
he before knew from feeling, he would carefully observe,
that he might know them again ; but having too many
objects to learn at once, he forgot many of them ; and
(as he said) at first he learned to know and again forgot
a thousand things in a day.   More than two months
elapsed after the operation, before he could be made to
comprehend that pictures represented solid bodies, when
to that time he considered them only as party-colored·
planes, or surfaces diversified with a variety of paint ;
but even then he was no less surprised, expecting the
pictures would feel like the things they represented, and
was amazed when he found those parts, which by their

light and shadow seemed round and uneven, felt only flat like the rest, and asked which was the lying sense, feeling, or seeing?

" Being shown his father's picture in a locket at his mother's watch, and told what it was, he acknowledged a likeness, but was vastly surprised ; asking how it could be, that a large face could be expressed in so little room ; saying, it should have seemed as impossible to him, as to put a bushel of any thing into a pint.

" At first, he could bear but very little light, and the things he saw he thought extremely large ; but upon seeing things larger, those first seen he conceived less, never being able to imagine any lines beyond the bounds he saw. And now being lately couched of his other eye, (more than a year after the first operation), he says, that objects at first appeared large to this eye, but not so large as they did at first to the other ; and looking upon the same object with both eyes, he thought it looked about twice as large as with the first couched eye only, but not double, that we can any ways discover."

When we began the study of vision it was _{Parts protect-ing the eye.} with the remark, that the apparatus charged with the exercise of this sense was composed of an essential part, the globe of the eye and optic nerve, and of various accessory parts destined to move or to protect the former. Enough has been said upon the structure and functions of the eye itself, and also of the muscles which are fixed to it: to close the history of this function, we have only to describe the protecting parts of this organ.

The first objects of interest are the bony cavi- _{Orbit.} ties, which lodge the eyes and are called the‾ *orbits*. These are deep holes hollowed in the face and bounded

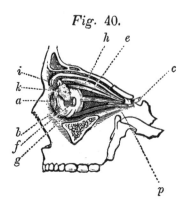

Fig. 40.

by the several bones of the head. They are very large, and have nearly the form of a cone, with the base directed outwards, and apex towards the brain, which is pierced by a hole for the passage of the optic nerve. In man and monkeys the orbits are directed forward, and their external wall separates them completely from the temporal fossæ; but if we examine animals gradually differing in an increased ratio from these, the orbits will be found to become more and more lateral, and to be more intimately confounded with the above mentioned fossæ.

But the globe of the eye is separated from the bony walls of the orbit by its muscles, and by a great quantity of fatty cellular tissue which surrounds it, as an elastic couch.

In front, the eye is protected by the eyebrows, eyelids, and by a peculiar liquid, the tears, with which its surface is always bathed.

Eyebrows. The eyebrows are transverse eminences formed by the skin, which in this point is supplied with hair and provided with a special muscle of motion. They serve to protect the eye against external violence, to prevent the perspiration flowing from the forehead from irritating the surface of this organ, lastly, to protect it from the impression of too strong a light, especially when the latter comes from an elevated place.

Eyelids. The eyelids in man, and all other mammiferæ,

are two in number situated one above the other, and for this reason divided into superior and inferior. They are a kind of movable veil placed in front of the orbit and accommodated in form to the globe of the eye, so that being closed they completely cover the anterior face of this organ. Exteriorly, they are formed by the skin which at this place is very fine, semi-transparent, and sustained by a fibro-cartilaginous plate (the tarsal cartilages). Their internal face is lined by a mucous membrane, called the *conjunctiva*, which is reflected upon the globe of the eye, covers the whole anterior part of the sclerotica, and is confounded with the transparent cornea. The free border of the eyelids is supplied by a range of hairs, having behind them a series of holes which communicate with the *glands of Meibomius*, follicles lodged within the tarsal cartilages and secreting a peculiar humor, which, when thickened and dried, as often happens in sleep, is known as the *wax*. Finally, there are within the eyelids muscles of motion; one of them surrounds the opening like a ring, and contracts them with more or less force (fig. 32, *h*); the other extends from the superior eyelid to the bottom of the orbit, and serves to raise this veil, (fig. 40, *i*).

The eyelids serve to prevent the access of light to the eye during sleep. When awake, they approach or separate so as to allow only the quantity of light necessary to vision to pass, but not enough to injure the retina. They also protect the eye from the contact of foreign bodies which are in motion in the air, preserve it from shocks, by their almost instantaneous occlusion, and oppose the effects of the prolonged contact of the air by continual motions, which return at nearly regular intervals.

One of the uses of the conjunctiva is to facilitate   Winking.

the operation of *winking*. This membrane, with an exquisite degree of sensibility, secretes a humor which
increases the polish of its surface, and renders the constant friction of the pallebral, upon the occular portion of
the conjunctiva, much more easy ; but this liquid is not
alone sufficient for this effect, and in order that the conjunctiva may properly discharge all its functions, its surface must be lubricated by the *tears*.

Tears.   This humor, composed of water, holding in
solution some thousandths of animal matter, and of salts
met with in all the liquids of the animal economy, is
formed in a large gland, situated in the vault of the orbit,
behind the external part of the border of this cavity and
above the globe, (fig. 40, *k*).

Lachrymal apparatus.   This *lachrymal gland* sheds the tears upon the
surface of the conjunctiva by six or seven small canals,
which open upon this membrane, on the superior and external part of the upper lid.   The tears afterward spread
over the whole surface of the conjunctiva, preventing its
desiccation and forming a uniform layer, which gives to
the eye its polish and brilliancy.   They must also serve
to prevent the evaporation of the humors of the globe of
the eye, and that of the liquids moistening the cornea ;
and it is actually the case, that when after death tears
cease to be diffused upon the surface of the eye, the latter
soon becomes flaccid, and the cornea loses its transparency.

The tears, which do not evaporate or are not absorbed
by the conjunctiva, pass into the nasal fossæ by means of
canals, the openings of which may be seen in the free
border of either lid, near the internal angle of the eye, at
the point where these organs quit the globe of the eye to

pass to the *caruncula lachrymalis*, a projecting body of a rose color, principally formed by a collection of small follicles. These two openings, styled *lachrymal points*, are extremely narrow and communicate with very fine canals, which are lodged in the interior of the lids and directed inward to empty into the *nasal duct*. This latter duct extends from the internal angle of the eye to the inferior meatus of the nasal fossæ, and to reach it traverses a bony canal hollowed between the orbit and the nose.

In the ordinary condition, the absorption of tears by the lachrymal points takes place but very slowly ; but when the latter become very abundant and roll in the eyes, their passage into the nasal fossæ becomes so rapid, that the handkerchief is requisite at every instant. Sometimes, in certain lively emotions of the mind, for example, the secretion of tears becomes so very abundant that this liquid runs over the eyelids and falls upon the cheeks.

Weeping may, also, depend upon another cause. Sometimes there occurs an obstruction of the nasal canal which prevents the tears from reaching the nasal fossæ ; this liquid then flows upon the cheek and accumulates with so much rapidity in the superior part of the obstructed canal, that a considerable tumor may result, or even its walls be ruptured and a *fistula lachrymalis* formed.

The structure of the apparatus of vision and the mechanism of sight, are nearly the same in *Apparatus of vision in animals.* man and all the mammiferæ, as in birds, reptiles and fishes. The eye of some of the mollusca, such as the *poulpe*[1] also resembles ours a little ; but in most animals

---

[1] This term is applied to such of the mollusca, belonging to the class cephalopoda, as inhabit chambered shells ; the Nautilus, some of the polypi, etc., are examples.

of this class and in the arachnida, the crustacea and insects, these organs have scarcely any points of resemblance with the eyes of the superior animals.

OF THE INTELLECTUAL AND INSTINCTIVE FACULTIES.

We have already seen how, in order that man may have the knowledge of the impression produced by the bodies acting upon his senses, it is not alone sufficient that these parts be endowed with the faculty of feeling, and that the sensations of which they are the seat be conveyed to the brain by the intervention of the nerves : but that, in order to have the impression thus received by the brain become a sensation of which we are conscious, this organ must not remain passive, but must perceive and acquaint us with it. And in very many cases a multitude of impressions are received by the brain without any consciousness on our part. In sleep, for instance, we are ignorant of noise at our ears, or odors acting upon the organ of smell. When awake, custom also renders us insensible to a great number of these impressions, and when several strike us at the same time, we may even, sometimes, by the mere effort of volition, abstract ourselves from some of them, and only really perceive those to which we wish to direct our attention.

Perception. This faculty of *perception*, which, when excited and directed by the will, takes the name of *attention*, is one of the attributes of the brain, and exists only when this organ is in activity. The experiments, in which the two hemispheres of the brain have been removed in living animals, have already demonstrated, that the loss of these parts rendered these beings insensible to all impressions received by the organs of the senses, and when the brain is prevented from discharging its functions, either by compression or the administration of certain poisons to the animal, as opium, the same effect is produced. Different diseases of the brain, also, render it unsuited to perception, and when it has been for some time in a state of continual activity, it presents the same phenomena with all our other organs ; it becomes fatigued and cannot continue to discharge its functions, till it has remained, for a longer or shorter time, in repose.

The brain does not merely possess the faculty Memory. of perceiving sensations and thus producing what are called *ideas*, it is also capable of recalling ideas already acquired. This new faculty bears the title of *memory*, and it is independent of perception ; for, in certain cerebral affections, we find it completely lost without depriving the patient of the faculty of knowing what surrounds him. It is remarked that the memory preserves the clearest idea of the most lively impressions, and that this faculty, extremely developed in the early period of life, is weakened with the progress of age. In the aged, memory is sometimes entirely lost, and in the adult it is weaker than in the adolescent or child. Therefore, at a tender age, such knowledge is acquired as does not demand great reflection, such as the languages, history, the descriptive sciences, etc., exercise tends also to render it stronger.

But memory can hardly be considered a single faculty,

29

and a multitude of facts go to show that there are, so to speak, as many distinct memories as there are orders of different faculties of the mind.   There is a memory of words, of forms, of places, of music, etc., and it is very rare that a man possesses them all in the same degree ; in general one of these faculties predominates, and in certain cases of mental disease, one of them has been completely lost, while the other kinds of memory were not sensibly weakened.

Reasoning, etc.   Acquired ideas do not remain isolated in the mind : we possess the power of comparing them, of seizing upon their respective relations, of drawing from them conclusions, and, in a word, of exercising our judgment upon all we perceive.   The faculty of forming a succession of judgments which are chained to each other, or of reasoning, is one of the most precious attributes of man.   But the peculiar characteristic of human intelligence, and, that which permits him to acquire the prodigious development manifested by him in civilized nations, is the faculty of generalization, which consists in creating signs to represent ideas, to think by means of these signs, and to form abstract ideas.

Instincts.   Finally, it would appear that, upon the action of the encephalon depend the inclinations or instincts, which induce man and animals to execute certain determinate actions, and to prefer certain sensations to those of another order.

Relation between the development of the brain and the intellectual faculties.   It is remarked that an organ acts powerfully in proportion to its volume.   It might then be supposed, that there would be a certain relation between the development of the encephalon and that of the intellectual or instinctive faculties, which appear to have their seat in it; and when man is compared with other animals, his brain is generally found to be proportionally more voluminous. It is, also, remarked, that animals which display the most intelligence, monkeys for instance, have this organ very large, while in the more stupid, as fishes, it is always extremely small.

These facts have led to the inference that we might judge of the degree of intelligence of animals, and even of man himself, by the greater or less development of their brain; and to appreciate these differences, various methods have been resorted to, the most celebrated of which is the measure of the *facial angle,* proposed by Camper, a skilful Dutch naturalist.

These measures serve to point out the relation existing between the volume of the cranium (which is filled by the brain and cerebellum), and that of the face, and are taken in the following manner. A horizontal line (*c, d*), is drawn on a level with the auditory hole and floor of the nasal fossæ, so as to follow nearly the direction of the base of the cranium; a second line is then let fall upon this (*a, b*), which touches the most prominent point of the forehead and the extremity of the upper jaw. Now, it is evident that this latter will be inclined upon the former, and will form with it an angle so much the more acute, as the face is more developed and the forehead retreating, and that, consequently, the measure of the *facial angle* (for so the angle of which we have just been speaking is called), may indicate with sufficient accuracy the relation sought.

Facial angle.

Fig. 41.

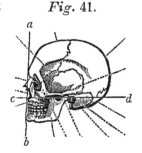

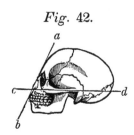

Fig. 42.

Man is of all animals the one with the facial angle most open, and in this respect there exist great differences among the different human races. European heads are usually about 80° (fig. 41), and negroes about 70° (fig. 42) ; in the different monkey tribes it varies from 65° to 30°, and becomes yet more acute as we depart from man and descend in the series of mammiferæ ; in the horse, for example, the forehead is so retreating that it becomes impossible to draw a straight line from the extremity of the superior jaw to the cranium, on account of the projection of the nose, as any one will soon be convinced by casting

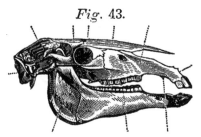

Fig. 43.

his eyes upon the annexed figure ; finally, in birds, reptiles and fishes, the facial angle, when it can be measured, is more acute than in the mammiferæ.

The greater or less coincidence which exists in general between the inclination of the facial line and the extent of the intellectual faculties, does not appear to have escaped the observation of the ancients ; not only have they truly remarked that the open angle was a sign of a more generous nature and one of the characteristics of beauty, but in the figures of their heroes and gods, they have advanced the facial line more than it is in man, and in some statues (that of Jupiter Olympus, for example,) they have inclined it a little forward.[1]

---

[1] It is possible, however, that this manner of representing the Divinity arose from another cause, and was independent of any idea of a relation

Most persons are accustomed to attribute stupidity to men and animals with retreating foreheads, or projecting snout ; and when by any circumstance the facial line is raised, even without augmenting the capacity of the cranium, we find, in animals which present this disposition, a peculiar air of intelligence, and we are induced to attribute to them qualities they do not actually possess. The elephant and owl are examples, and this is caused by the great extent of the frontal sinuses giving to their forehead a considerable prominence.   The owl, as every one knows, was with the ancients an emblem of wisdom, and the elephant bears in the Indies a name which signifies that he partakes of reason, and yet neither of these animals is actually remarkable for the development of its intellectual faculties.

However, we must beware of attaching too much importance to these measures ; they can give at the most but a proximate idea of the development of the brain, and as yet there is no proof that the extent of the intellectual faculties is proportional to this material development of the encephalon.

We have already seen that the brain is the _{System of Doctor Gall.}

between the development of intelligence and the opening of the facial angle. All people attach ideas of beauty to the exaggeration of peculiarities of structure characteristic of their race; negroes esteem the blackest skins the handsomest; the inhabitants of Oceania, with remarkably flat noses, think their beauty is increased by augmenting the width of this part; and the Caraibs, with extremely retreating foreheads, compress the heads of their children, to exaggerate yet more this characteristic disposition.   Now one of the peculiarities of the Cossack race, and more especially of the Greek nation, is the slight inclination of the facial line, and consequently from the tendency just observed the Greeks might naturally regard this disposition as a condition of beauty, and think that to represent beings superior to ourselves, it must be exaggerated.

seat of many very distinct functions, and when we examine the manner in which the intellectual faculties and
feelings are exercised in different men, we soon observe
that a more or less increased development of one is not
always accompanied by an equal development of the rest.
A man, who shall be remarkable for the instinctive love
to his offspring, may have but very feeble intellectual
faculties; and another, endowed with a most happy disposition for the sciences of calculation, may yet lack completely imagination, or power of observation.

These considerations and many similar facts, have led
some philosophers to consider the brain as a single organ,
all parts of which concur in the same manner to the production of the phenomena of instinct and intelligence,
but that nature has established in the functions of the
encephalon the same division of labor that we find in
the other apparatus of the animal economy, when the
faculties of the latter are perfected.   They have regarded
the feelings as having their seat in a determinate part of
the brain, the intellectual faculties in another, and, in a
word, each kind of function executed by the brain as
the result of the action of an instrument or particular
organ, and that these special organs were different portions of the nervous mass of the encephalon.

Upon this hypothesis of the localization of the various
functions of the encephalon is founded the *phrenological*
system of Dr. Gall.

He considers each of these functions to be the appendage of a determinate part of the brain or cerebellum, and
that the greater or less activity of each of them depends
in a great measure upon the development of the part
which is their seat.   Now in man and most of the supe-

rior animals, the encephalon fills the whole cavity of the cranium, and the walls of this osseous case are in some sort moulded upon the nervous mass, so that one may judge of the proportional size of the different parts of the brain by the prominence of the corresponding parts of the head. And admitting these suppositions to be correct, we may consequently judge by mere inspection of the cranium, of the inclinations and faculties of every individual.

Phrenologists admit that the feelings which give to animals their inclinations or excite their desires are situated in the posterior and inferior parts of the encephalon; the instinct of propagation resides therefore in the cerebellum; the love of offspring would depend upon that part of the third cerebral lobe which is seen immediately above this organ; the instinct which renders animals more or less social would result from the action of a neighboring part; courage would depend upon that part of the brain situated above and behind the ear; the love of destruction on that directly above the ears; finally, the inclination which induces to the employment of stratagem and the desire of acquisition would occupy neighboring parts. The feelings on which depend the sentiments of self-love, of vanity, circumspection, benevolence, firmness, justice, etc., would have their seat in the superior and anterior parts of the brain; lastly, the different intellectual faculties have been assigned to the parts corresponding to the forehead.

A great argument in favor of this hypothesis may be drawn from the peculiarities to be observed in the configuration of the heads of men, remarkable for certain qualities of mind, or the force of certain inclinations, and the variations observed in the form of the cranium of animals

of the most opposite instincts.   What has already been
said with regard to the facial line applies especially to the
more or less considerable development of the anterior
part of the brain, and the existence of a depressed and
retreating forehead will give the whole head an aspect of
stupidity.   It is also remarked, that in carnivorous animals living by the chace and which display the most
courage and ferocity, the width of the cranium, at the
ears, is much greater than in the herbivorous, with mild
and timid manners.   It must also be allowed that in almost
all animals the posterior part of the head, where phrenologists place the love of offspring, seems to be more developed in females than males, and we all know that the
tenderness of a mother for her young is a much stronger
passion, than in the father.

But if some of the suppositions which form the basis of
phrenology appear to be really quite plausible, others are
not founded upon any conviction and must even appear
absurd to all persons accustomed to analyze the phenomena of intelligence.   Thus some phrenologists admit a
particular faculty for the appreciation of weight, another
adapted to judge of length and so on.

But we repeat, no fact is yet known suitable to prove
that this division of labor really exists in the brain, and
some experiments of M. Flourens would even lead to the
idea that it does not.

MOTIONS.

The various modifications of the faculty of perception
which have been noticed in the preceding pages, render
man and animals capable of knowing surrounding objects;
but their relations with the external world do not merely

consist in these phenomena, in some sort passive. These beings may also act upon foreign bodies, make changes in them, move, and often express in a manner, more or less precise, their sentiments or ideas.

This new series of functions, with which we Contractility. are now to be occupied, depends essentially upon a property not less general among animals than sensibility, to wit, *contractility*.

This name is given to that faculty possessed by certain parts of the animal economy, of alternate contraction and extension.

In some animals of an extremely simple structure, Muscles. such as the hydræ, all parts of the body appear susceptible of this contraction ; and however little we ascend in the series of beings this faculty soon becomes the property of particular organs, called *muscles*. These muscles, which are the active instruments of all our motions, form the greater part of the mass of the body and constitute what is commonly called the meat, or flesh of animals. Their color is generally whitish. In some animals they are, on the contrary, of a more or less intense red ; this color does not necessarily belong to them, but simply depends on the blood they contain.

Each muscle is formed by the union of a certain number of muscular bundles, which are united by cellular tissue and composed of smaller bundles. These in their turn are formed of bundles less in volume, and, from division to division, we finally arrive at fibres of an extreme tenuity, which are straight, ranged parallel-wise, and which, seen by a powerful microscope, appear to be formed each by a series of small globules of about one three hundredth of a millemetre in diameter. After death the muscular tissue is

30

soft and easily torn, but during life it is very elastic and resisting.   Finally, it is essentially composed of a material already met with in the blood and called by chemists *fibrine.*   We also find albumen, osmazome and some salts.

Muscular
contractions.   Under the influence of certain exciting causes, the muscular fibres are contracted, and at the same time the bundles they form become larger and harder than in the state of relaxation.   Any one can observe this phenomenon upon himself, if he executes any movement and observes the changes which take place in the muscles called into action in producing it.   Let any one forcibly bend the fore-arm upon the arm, for example, the muscles of the anterior part of the arm will be found swollen and hardened.

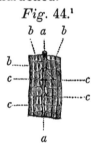

Fig. 44.[1]

By the aid of the microscope we can easily distinguish the manner in which this contraction is made.   When the muscular fibres are in a state of relaxation, they are extended in a straight line (fig. 44); but when they contract, they suddenly take a zigzag course and present many angular and regularly opposite undulations (fig. 45).   By repeating this experiment we are soon able to recognise that the flexions of each fibre take place in certain determinate points, and never otherwise.   When

---

[1] Fig. 44. Portion of a muscle, in the state of repose, seen by the microscope to demonstrate the disposition of the fasciculi of muscular fibres and the mode of distribution of the nervous filaments, — *a*, nerve; *b b*, fasciculi of muscular fibres arranged parallel and in a straight line; — *c*, nervous filaments which separate from the nerve *a*, and traverse perpendicularly the muscular fasciculi at equal distances.

the contraction is feeble, these flexions are slightly marked, and in the stronger contractions they never advance beyond angles of 50°.

Thus by contraction the two extremities of the fibre are approximated without in any respect changing its total length. Now, these extremities are fixed to the parts the muscles are to move, and by the displacement of the former, the latter are drawn with them.

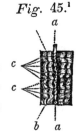

Fig. 45.[1]

This insertion of the muscles into movable parts is not made directly, but takes place by means of an intermediate substance of a fibrous texture, which penetrates into the substance of these organs, so as to send a prolongation to each of the fibres composing them. Sometimes this fibrous tissue, which is white and pearly, takes the form of a membrane and then it is called an *aponeurosis;* at other times it resembles a cord of varying length and then constitutes what anatomists call *tendons.*[2]

We said above that contractility appertained   Influence of the nerves. exclusively to the muscular fibres : these are actually the only parts of the animal economy which possess the faculty of contraction ; but this property they owe to the nervous system.

Each muscular bundle receives one or several nerves. These nerves, which are surrounded by a sheath called *neurilemma,* are composed of many longitudinal filaments and these filaments are scattered throughout the whole

---

[1] The same muscle at the moment of contraction ; the letters *a, b, c,* indicate the same parts as in the preceding figure.

[2] The tendons and ligaments are vulgarly called the nerves, although they have nothing in common.

muscle, passing nearly parallel between and transversely upon the muscular fibres, precisely in the points corresponding to each of the angles, formed by the zig-zag folds, upon which contraction depends. After having thus continued their course during a certain time, these nervous fibres are seen to curve, bend, and return to the brain, so that they appear to form with this organ a continuous circle.

Now, if a nerve, which is thus distributed to the muscles, be divided, and thus separated in a manner from the central mass of the nervous system, its fibres are prevented from contraction ; they are paralyzed. It is sufficient, merely to compress the brain in a living animal, to cause it to lose the faculty of executing motions.

Galvanic experiments. Many researches have been made to ascertain the nature of the influence exercised by the nervous system upon the muscles, when it determines their contraction. The most celebrated are those of a physician of, Bologna, Galvani ; and at the same time that they have thrown new light upon this delicate question, they have conducted to one of the greatest discoveries of the past century, that of galvanic electricity.

The labors of Galvani, Volta, and of some other distinguished men, have demonstrated that when certain bodies of different natures, copper and iron, for example, touch, they develope *electricity,* and this electricity passes with great rapidity through certain bodies, such as the nerves and the metals, which for this reason are called good conductors of electricity, while it is arrested by others, such as glass and resin.

When a muscle is paralyzed by the section of the nerve going to it, we may, during some time, supply the want of nervous action by electricity ; and by means of this agent cause contractions similar to those which, under ordinary circumstances, take place from the influence of the will.

The most convenient manner of making these experiments is to deprive a frog of its skin, and divide the animal on a level with the loins, then to seize the lumbar nerves and envelop them in a small sheet of tin foil ; then the abdominal members are to be placed upon a plate of copper, and every time that the foil touches this latter metal, the muscles are seen to contract, the legs bend and are agitated, and this half of the frog seems to be leaping, as when alive. These singular effects may be produced sometime after the death of the animal, and are also observed in man ; for by passing an electic current through the bodies of some criminals, they were thrown into violent convulsions.

An analogous phenomenon takes place, when upon dividing a nerve in a living animal, the portion adherent to the muscles is pinched or burned, the latter immediately contract, but yet this effect appears to depend upon the same cause with the convulsions produced in the preceding experiments, for it has been proved that in all these cases there is a production of electricity.

From the preceding statements we learn that *Theory of muscular contraction.* the electric currents act upon the muscles in the same manner as the nervous influence, and the knowledge of this fact has led to a very plausible explanation of muscular contraction.

Physics teach us that when an electric current traverses, in a contrary direction, two parallel branches of a conducting rod, an iron wire, for example, they will be found to approach each other.

We have also seen that the nervous fibres in their distribution to the muscles, form elbows. It must then be
supposed that the electric fluid, which has traversed them
in all the preceding experiments, follows the course of
these curved fibres, and consequently descends by one
branch to ascend by the other parallel branch. If this be
so, these nervous fibres must be found in the same condition with the metallic conductor just mentioned; they
must approximate, and by approximating, they must draw
with them and fold the muscular fibres they traverse.

Now, the same phenomena may be observed in the
muscular contractions produced by means of the nervous
influence, and in those caused by electricity; it may
therefore be supposed that in the two cases, the cause
which produces them is, if not the same, at least very
analogous, and that in the normal state these contractions
depend upon the passage of a fluid having, in this respect,
the same properties with the electric fluid.

In this hypothesis, which is due to Messrs. Prevost
and Dumas, the muscular fibres would be mere passive
instruments in the phenomena of contraction, and these
nervous turns would be the true motor agents. A circumstance in support of this opinion, which has been
established by many delicate experiments, is the constant
relation already mentioned between the point in which
these nervous fibres traverse the muscular fibres, and
that in which the latter bend in contraction; the nerves
are always found at the summit of the angles formed by
these folds, and this is really the place they should
occupy, if their approximation were the cause of these
curves.

However, we see that contraction can take place only

in the muscular tissue, and that the action of the nervous system is its determining cause. Let us now inquire what are the parts taken by the several portions of this system, in the production of this important phenomenon.

The muscles present in themselves very great differences; some contract only under the influence of volition, others are equally under the empire of this force, but their contraction may also take place independently of it; finally, there are yet others on the motions of which volition has no influence. The muscles of the limbs, etc., belong to the former of these three classes, those of the respiratory apparatus to the second, and the heart, etc., to the third.

All those muscles whose motions can be caused by volition receive nerves from the cerebro-spinal system. But all the nerves of this system do not discharge these functions; some, as we have already seen, belong exclusively to sensation. The cerebral nerves of the third, fourth, sixth, seventh, ninth, and eleventh pairs, (fig. 29) appear, on the contrary, to be exclusively assigned to motion; finally, the cerebral nerves of the fifth and tenth pairs, and all the nerves which arise from the spinal marrow, discharge these functions at the same time that they serve for sensation. Their anterior root, as already seen, gives them the faculty of transmitting sensations to the brain; and by their posterior the nervous influence necessary for voluntary motion is propagated from the brain to the muscles. *Nerves of voluntary motion.*

In fact, when the posterior roots of the spinal nerves are divided in the living animal, the parts to which they are distributed are deprived of the power of contraction, exactly as if both roots had been divided.

Functions of the spinal marrow.   When the spinal marrow is divided, the movements of all the parts whose nerves originate below the section are suspended; while those whose nerves are yet in communication with the brain, continue Brain. to be moved. But if instead of thus experimenting upon the spinal marrow, we act upon the brain, by removing or compressing it so as to prevent the discharge of its functions, all the muscles of voluntary motion are paralyzed at the same time.

Of the striated bodies.   It would also appear that certain parts of the nervous system exercise a different influence over the motions. Thus M. Magendie has proved that if the portion of the brain, called by anatomists the *corpora striata*, be cut into, the animal thus mutilated is no longer master of his movements, but seems driven forward by some internal force which he cannot resist; he leaps forward, runs with rapidity, and finally stops, but seems unable to Cerebellum. move backward. If, on the contrary, the two sides of the cerebellum be wounded in one of the mammiferæ or in a bird,[1] it will walk, swim, or even fly backward, without being able to advance.

When these lesions are only practised on one side, other phenomena will be observed, which at first sight appear much more singular, but which are consequences of the effects already mentioned. Thus, if one side of the cerebellum or of the annular protuberance be cut vertically, the animal begins at once to roll on its side, turning from the wounded part, and sometimes with such rapidity, that it makes more than sixty revolutions in a minute.

---

[1] From the experiments of M. Magendie it would appear that the same effects were not observed in reptiles and fishes.

From these curious experiments, and the researches upon the same subject by Flourens and some other physiologists, the cerebellum and neighboring parts of the encephalon have been found to possess, among other properties, that of regulating the movements of locomotion.

The movements which, although under the direction of the volition, are also independent of its influence appear then to depend upon the action of the medulla oblongata. In fact, when the brain no longer discharges its functions, and when consequently there can be no volition, the muscles of the respiratory apparatus continue to act as if their movements were under the direction of the will; but when this portion of the medulla is destroyed, although the brain be left intact, they are immediately arrested.

Respiratory movements.

Functions of the medulla oblongata.

Those muscles, whose contractions are entirely independent of the will, receive their nerves from the ganglionary system, and in this system resides their principle of action ; for if respiration be maintained by artificial means, the whole encephalon may be destroyed, as well as the spinal marrow, without arresting the beating of the heart, or the peristaltic motion of the intestines.

Involuntary motions.

The contraction of the muscular fibre is a phenomenon essentially intermittent. The muscles cannot remain in a state of permanent contraction, and at the end of a longer or shorter time they are necessarily relaxed. Thus the heart, whose action only ceases with life, alternately contracts and reposes; but in the muscles of voluntary motion, these same contractions, interrupted by longer or shorter intervals of repose, cannot be continued beyond a certain time, for they produce

Laws of muscular contraction.

31

a feeling of lassitude which increases till at last these movements become impossible, which is only relieved by a period of inaction or repose.

The ease with which muscular fatigue is manifested varies greatly according to the individuals ; but, cæteris paribus, it is in the ratio of the intensity of the contractions, the duration of each of them, and the rapidity with which they succeed each other.

The force displayed by the contraction of a muscle depends upon the texture of this organ ; and the nervous energy of the individual. The large, firm, and red muscles, are capable of contracting with more power than the small, flabby, and pale ; but only when these conditions are united to a very strong volition, can these organs produce the greatest effects, and almost always they are in an inverse ratio. The energy of the muscular contractions may be carried to an extraordinary degree by the sole influence of the action of the brain. We acknowledge the power of an angry man and of a lunatic ; and when, in the ordinary state of the economy, a similar nervous energy is united to a great material development of the muscular system, astonishing effects result ; of such the ancients have transmitted us recitals in speaking of their athletæ, and the buffoons of our own day are also sometimes examples.

Muscular contraction is an important office in several of the functions, the history of which has already been given ; but the subject to which we shall now turn our attention is more directly connected with it, for we are about to enter upon the study of the general and partial movements of the body, upon which depend attitudes, locomotion, and many other phenomena entirely mechanical.

In the inferior animals the muscles are all in- ^{Passive organs of motion.} serted into the tegumentary membrane, which is soft and flexible; and by acting upon it they so modify the form of the body as to impress on it motion entirely or in part; but in animals of a more perfect structure, the apparatus of motion is more complicated, and is composed not merely of muscles, but also of a system of solid pieces which serve to increase the precision, force and extent of the motions, at the same time that it determines the general form of the body and protects the viscera against external violence.

This solid frame, to which the muscles are at- Skeleton. tached, bears the name of *skeleton*. In certain animals, such as insects and lobsters, it is situated externally, and consists merely in a modification of the skin; but in man and all animals nearly allied (namely, the other mammiferæ, birds, reptiles and fishes,) it is situated in the interior of the body, and is composed of parts peculiar to itself.

In some fishes (the rays, for instance,) the skel- Cartilages. eton is formed of a white substance, opaline, compact, to appearance homogeneous, very resistant and elastic, which is called *cartilage*. It is the same with the skeleton of man and other animals at the early period of life; but this state, which is permanent in the fishes spoken of, is in the other animals but transitory, and the cartilages of the skeleton become encrusted with stony materials of a calcareous nature, which render them stiff, brittle, and very hard, and which transform them to the state Bone. of *bone*.

To be convinced that bones are but cartilages, hardened by the deposition of calcareous salts into their substance, it is only necessary to macerate them for some time in a particular liquid, called muriatic or hydrochloric acid. This liquid has the property of dissolving the stony materials contained in the bones without attacking the cartilages, so that the latter may thus be separated from the salts which masked its properties.[1]

Development of the bones. The ossification of the skeleton commences by an infinity of points which are constantly extending; therefore the number of distinct osseous pieces is at first immense; but by the progress of ossification several among them are united, so that in the adult there are fewer distinct bones than in the infant; and in extreme old age, several bones are often united in one, and parts originally cartilaginous are encrusted with calcareous matter. The utility of this mode of development can be at once comprehended; in order that the solid frame of the body may not oppose its motions, the former must always be composed of a great number of movable pieces, but this division is especially necessary when all these parts must yield to the increase of the organs situated in its interior.

Structure of the bones. The surface of the bones is always covered by a membranous layer, to which is given the name of *periosteum*, and their substance is composed of fibres or

[1] From the analysis of M. Berzelius the bones of the human skeleton, completely deprived of fat, are composed to the hundred parts, as follows: of cartilage, 32.17; vessels, 1.13; neutral phosphate of lime, with a small portion of the fluoride of calcium, 53.04; carbonate of lime, 11.30; phosphate of magnesia, 1.16; soda, with a small proportion of chloride of sodium, 1.20. In the bones of the ox this chemist found the same proportion of animal matter, but less of the carbonate of lime. The cartilaginous part of the bones is composed of gelatine, and is therefore used in the arts, and in domestic economy, for the manufacture of glue and the preparation of economical broths.

lamellæ, easily distinguished. When these organs are to occupy a small space and present great solidity, as is the case with the flat bones, which cover the majority of the most important and delicate viscera, the osseous tissue is extremely compact. But when the bones are to occupy a long space, the motions would be injured if their weight were considerable, their tissue is dense and close only upon the surface, and in their interior exist large cellules or even canals, called medullary, because they are filled with marrow.

The bones vary greatly in form, and from this Form. variation they are divided into long, short and flat bones. The former simply present a medullary cavity and are always nearly cylindrical. We often find in all of them eminences which give attachment to the muscles or to other parts, and which, when they form a considerable prominence, are designated as *processes*. The bones also present upon their surface depressions more or less shallow, and which serve to lodge the soft parts or to receive other bones which are to move in these cavities, and in many places, are holes which afford a passage to the blood vessels or nerves.

The name of *articulation* is given to the union Articulation. of the various bones together. The means of conjunction employed by nature for this purpose vary much, according as the bones are always to preserve the same relation and remain fixed, or to execute motions of greater or less extent.

When the articulation of the bones is not destined to permit motion, it may take place in three ways : by juxtaposition, by suture, or by implantation. The articulations by simple juxtaposition of the articular surfaces are only seen in certain parts of the skeleton, where the position of the bones is such that they cannot be displaced. In the articulations by suture, the articulating surfaces present a series of asperities and angular projections, which are reciprocally received; thus these articulations may have great solidity with but small extent of surface. Finally the articulations by implantation are those, in which a bone is set into a cavity hollowed in the substance of the bone which serves as its base: these are the most solid but are rare.[1]

In the movable articulations, the bones are not directly united together, but are maintained in contact by bonds extending from one of these bones to the other.

Sometimes these articulating surfaces are united by an intervening cartilaginous or fibro-cartilaginous substance, strongly adherent to both of them, and which only allows them to move by reason of its elasticity, (this is called the *articulation by continuity*): at other times the articulating surfaces slide upon each other, and are only kept in contact by *ligaments*,[2] which surround them and are so arranged as to limit their motions. The latter mode of conjunction constitutes what anatomists call *articulation by contiguity*, and is met with in all cases of very extended motion. The surfaces thus articulating are always extremely smooth and surrounded by a cartilaginous layer, which highly increases their polish; but these are not the only means employed by nature to diminish the friction in the joints: for she has placed in them a sort of mem-

---

[1] The teeth, which are not true bones, are the only parts thus articulated.

[2] The name of *ligaments* is given to collections of fibres similar to those of the tendons, very resisting, round or flat, and of a pearly white, which fasten together the bones.

branous pocket, called the *synovial sac,* which is similar to the serous membranes, and is filled with a viscid liquid, which permits these surfaces to slide readily upon each other.

All the muscles destined to produce the great movements of the body, are fixed to the skeleton by their two extremities. Therefore, by contraction, they must displace the bone which presents to them the most feeble resistance, and draw it towards the one which remains immovable, and which serves as a fixed point to move the former. Now, in the majority of instances, the bones are more movable as they are at a greater distance from the central part of the body ; consequently, the muscles which are fixed to two of them act in general upon the more distant, and the muscles destined to move a bone always extend from this organ to the trunk. Thus the muscles serving to bend the fingers occupy the palm of the hand and the fore-arm ; those which bend the fore-arm upon the arm occupy the arm, and those which move the arm upon the shoulder occupy the shoulder.

In certain circumstances, however, these muscles displace the bones which in ordinary cases answer the purpose of a fixed point. When we attempt to raise the body hanging by the hands, the flexor muscles of the fore-arm not being able to displace the latter, approximate the arm to it, and thus draw up the whole body.

The kind of movement caused by the contraction of a muscle generally depends on the one hand, upon the nature of the articulation of the bone displaced, and on the other, upon the position of the muscle with regard to this bone : it always draws it to its own side, and approxi-

mates it to the point where its opposite extremity is fixed.
Thus the muscles which flex the fingers occupy the pal-
mar face of the hand and fore-arm, while those for their
extension are upon the opposite side of the limb.

Several muscles are often so arranged as to concur in
the production of the same motion.   They are then said
to be *allied ;* and the *antagonist* of a muscle is the one
which causes the opposite motion.

The muscles are also designated, from their uses, as
flexors and extensors, adductors and abductors, rotators,
etc.

Force of the muscles   The force with which a muscle contracts depends
upon its volume, the energy of the will, and some other
circumstances already mentioned ; but the effect produced
by this contraction depends, also in a great measure, upon
the manner in which it is fixed to the bone to be moved.

Thus, all things being equal, the motion caused by the
contraction of a muscle will be more powerful in proportion as the muscle is less obliquely inserted upon the bone
to be moved : when inserted at a right angle its whole
force is employed in displacing the latter ; but in the opposite case, a more or less considerable part of its force is
lost.

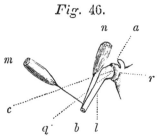

Fig. 46.

Thus if the muscle *m*, the
force of which we suppose
equal to ten, is fixed perpendicularly to the bone *l*, the extremity of which *a*, is movable
upon the fulcrum point *r*, it will
only have to overcome the
weight of this bone, and will carry it from the position *a b*,
in the direction of the line *a c*, causing the point of its

insertion to describe a space which we will also represent by 10. But if this muscle acted obliquely upon the bone, in the direction of the line *n b*, for example, it would then be quite different; for it will tend to carry it in the direction *b n*, and consequently to approximate it to the articular surface *r*, upon which the extremity of the bone reposes, but the latter being an inflexible stem this displacement cannot take place; the bone can only turn upon this point *r*, as on a pivot, and the contraction of the muscle *n*, without any loss of the energy we have attributed to it will only be able to carry this bone in the direction *a d*, and consequently to produce a displacement for which one fourth of the force would have been sufficient in its first position perpendicular to the bone.

The muscles are inserted for the most part very obliquely into the bones, and consequently in a manner very little favorable to the intensity of the result of their contraction. There often exists, however, a disposition which tends to diminish the obliquity of *Fig.* 47. *Fig.* 48. these insertions : this is the swelling met with at the extremities of most of the long bones, and which principally serves to give their articulations greater solidity. The tendons (*i*), of the muscles (*m*) situated above the articulation are generally inserted directly below this swelling, and thus arrive at the movable bone (*o*), by following a direction more nearly approaching the perpendicular, as any one may convince himself by comparing the disposition of muscle *m* in figure 48, where these swellings exist, with figure 47, in which the articulating extremities have been represented without any such swelling.

32

The distance separating the point of attachment of the muscle from the fixed point on which the bone moves, and from the opposite extremity of the lever represented by this organ, also exerts a very great influence upon the effects produced by its contraction. To explain this fact we must have recourse to Mechanics.

Levers. The bones we say represent *levers*, a name given in physics to every inflexible rod which moves upon a fixed point, called the *fulcrum*. The force which puts the lever in motion is called the *power*, and that which opposes its displacement is called the *resistance*. The distance which separates the fulcrum point from the point to which the power or resistance is applied, is called the *arm of the lever*.

The length of the arm exerts a very great influence upon the force necessary to constitute an equilibrium to a given resistance. For proof let us observe the mechanism of the *Roman balance*, as it is called. The beam

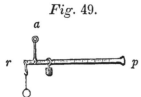

Fig. 49.

is divided into two parts of unequal length by the fulcrum (*a*). At the extremity of one of the branches (*r*), which is very short, is found the resistance (or object to be weighed), and upon the other (*p*) slides a weight, which makes the equilibrium to a resistance so much the more considerable as it is farther removed from the fulcrum and consequently the arm of the power of the lever lengthened, that of the resistance remaining always the same.

Every one knows too how great the difference is in the power a man can employ when he seeks to raise any burden with his arm bent or extended. In these motions, the same muscles are called into action, and the lever-arm of the power remains the same, it is only the lever-arm of the resistance, represented by the distance from the shoulder to the hand, which is lengthened.

The science of mechanics teaches us, that to constitute an equilibrium on any lever whatever, the resistance and the power must be proportional to the length of the arms of the lever, that is, when multiplied by their respective arms, they must both give the same product.

Thus, to make an equilibrium to a resistance (r), equal to 10, applied to the extremity of a lever (a b) of the length 20, the power

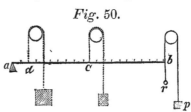

Fig. 50.

(p), if applied at the same point and consequently at an equal distance from the fulcrum (a), must also be equal to 10.   But if applied at the point c, to produce the same effect, the power must be equal to 20, for the resistance, which we have supposed equal to 10, being multiplied by the length of its arm of the lever (20), will give as a product 200, and on the other hand, the power arm of the lever (c a) being only equal to 10, the latter must be multiplied by a force equal to 20, to give the same product 200.   Finally, if the power be placed yet nearer the fulcrum, at the point d, it must have a force equal to 100; for its lever arm will be only 2, and $2 \times 100 = 200$.

The disposition of the levers exerts as great influence over the rapidity of the motions produced, as over their force ; and if, by employing a power comparatively feeble, a much greater resistance can thus be overcome, a slower or more rapid movement may also be obtained with the

aid of these instruments, by the employment of a motor
force of a certain rapidity.

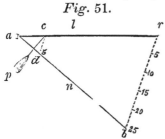

Fig. 51.

Thus let us suppose the
power *p*, to act upon the
lever *a r*, so as to cause the
point of insertion *c*, to describe a space of 5 in a
second, it will displace at
the same time the extremity
r, of the lever and cause it to arrive at *b*, with a rapidity
equal to 25, for the distance described at equal periods
by this point will be five times as great as that described
by the point *c*. The application, then, of a force whose
rapidity is only 5 to the point *c*, produces the same result
as if there had been directly applied to the point *r*, a
force whose rapidity is equal to 25.

But from what has been already said we see, that whatever is gained in rapidity is lost in force, for such results
are obtained by making the resistance arm of the lever
longer than that of the power.

Now, in the animal economy, nearly all the levers
represented by the bones are so arranged, as thus to
favor the rapidity of motion at the expense of the force
necessary to produce them. Thus when the extended
arm is brought down to the side, if the rapidity with
which the muscles contract is such that their insertion
be displaced three inches in a second, the extremity of
the limb will be removed, from its primitive position,
with the rapidity of nearly three feet in a second.

Having made these preliminary remarks upon animal
mechanics, we may now undertake the study of the
different parts of the apparatus of motion, which we shall
first examine in man.

The skeleton, as we have already said, is Skeleton.
composed of a great number of bones united together ;
it is divided like the body, into three parts, the head,
trunk and limbs.

The most important part of the skeleton, which serves
for the support of all the others, and differs least in the
different animals, is the *vertebral* or *spinal column.*

This name is given to a kind of osseous stem extending the whole length of the trunk, and which is composed of a great many small bones called vertebræ, which
are placed one upon the other and solidly united together.

This column (fig. 27), also called the *spine*, Vertebral column.
occupies the median and posterior line of the *Fig. 27.*
body, and supports on its anterior extremity
the head, which may be considered as its continuation. In man we reckon thirty-three vertebræ, and they are divided into five portions,
viz. : a cervical portion composed of seven vertebræ, a dorsal of twelve, a lumbar of five, a
sacral also of five, and a coccygeal of four. It
presents several curves and increases in size
from its anterior or superior extremity to the
commencement of the sacral portion. At birth,
all the vertebræ are perfectly distinct, and are simply
articulated with each other ; but in a short time the five
sacral are united and form but one bone, called the
sacrum (*s*).

The essential character of the vertebræ is *Fig.* 28.
to be traversed by a hole, which, by uniting
with those of the other vertebræ, forms a canal,
extending from the cranium to the extremity
of the body, in which the spinal marrow is

Fig. 52. — SKELETON OF MAN.

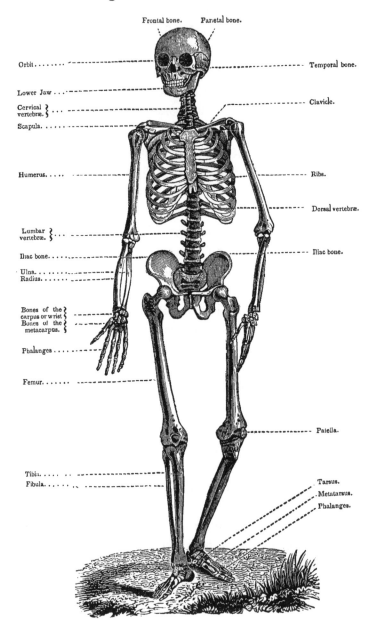

Frontal bone.   Parietal bone.

Orbit . . . .

Lower Jaw . . . .

Cervical vertebræ. . . .

Scapula. . . . . . .

Humerus. . . . . .

Lumbar vertebræ. . . .

Iliac bone. . . . . . .

Ulna. . . . . . .

Radius. . . . . . .

Bones of the carpus or wrist

Bones of the metacarpus.

Phalanges . . . . .

Femur. . . . . . .

Tibia. . . . . . .

Fibula. . . . . . .

Temporal bone.

Clavicle.

Ribs.

Dorsal vertebræ.

Iliac bone.

Patella.

Tarsus.

Metatarsus.

Phalanges.

lodged. In man the vertebræ of the coccyx present no such canal, for they are reduced to a rudimentary state and merely consist of so many solid nuclei. On the sides, this vertebral canal communicates externally by a series of holes, called *holes of conjugation*, because they result from the union of two grooves in the superior and inferior borders of each vertebra, so as to correspond when these bones are united. These holes, as already shown, give passage to different nerves which arise from the spinal marrow, and are distributed to the several parts of the body.

Each vertebra is divided into a body and several processes. The *body of the vertebra* (fig. 28, *a*,) is a thick disc, situated in front of the vertebral canal (or below, if the column is in a horizontal position, as in most animals), and serving to give solidity to the articulations of these bones with each other. The two faces of this disc are nearly parallel, and each of them is united to the corresponding surface of the neighboring vertebra, by a thick layer of fibro-cartilage which adheres to each, over the whole extent of the articulating surfaces, and only permits them to be separated so far as its elasticity will allow. The articulation of the vertebræ is also strengthened by the existence of four small processes, situated on the sides of the vertebral canal, and which interlock with those of the neighboring vertebræ. Finally, behind this canal, there exists a process called spinous (*b*), which assists in the same purpose, by limiting the flexion of the column backward; and fasciculi of aponeurotic fibres extend, also, from one bone to the other so as to fasten them together.

The articulation of the vertebræ is thus rendered extremely solid, and the motions of each of these bones are in general very limited; but the addition of these slight motions to each other, gives sufficient flexibility to the whole column without injuring its strength. But this mobility varies greatly in the different parts of the spine; in the lower part it is almost null, in the loins, on the contrary, it is quite marked, but it is the greatest in the cervical portion of the column; therefore, in these parts, the fibro-cartilaginous layer, in order to allow of these movements, is thicker than in the back, and the spinous processes are farther removed from each other, so as to allow of a considerable curvature of the column before they meet.

The weight of the body tends constantly to curve the vertebral column forward. To resist this flexion and straighten the column, powerful muscles were requisite, which are inserted the whole length of its posterior face; and to render their action more powerful, nature has so arranged their point of attachment as to allow them to draw perpendicularly upon quite a long arm of a lever. In fact, most of them are fixed to the extremity of the processes called spinous, which form a prominent crest the whole length of the spine, and others take their origin from two other processes (c), which are equally prominent and are named from their direction, transverse processes.

It must, also, be observed that in the portions of the column where these muscles are to employ the most power, as in the loins, these processes are much longer, and consequently form a lever far more powerful than in the parts where all this force is not necessary, as in the neck, for example. In animals with a heavy head, situated at the extremity of a long and horizontal neck, these processes have an extreme length on the back, where they serve for the attachment of the ligaments and muscles intended to sustain these parts and support the neck.

The motions of flexion of the column forwards require scarcely any force, and the muscles employed to produce them, and situated in front of the body of the vertebræ, are, consequently, small and few in number.

The first vertebra of the neck, called the *atlas*, is much more movable than any of the rest; it has the form of a simple ring, and turns upon a kind of pivot formed by a process which ascends from the body of the next vertebra (or *axis*). Upon this articulation are effected nearly all the motions of rotation of the head. The bonds which unite these two vertebræ are far less strong than those of the other vertebræ; and in fact, in the ordinary position of the body, the weight of the head pressing upon the atlas, tends rather to maintain them in contact than to separate them. When, however, the head supports the entire weight of the body, as in those who are hung, the case is altered; these two vertebræ then separate easily, and their luxation produces almost instantaneous death in consequence of the compression of the spinal marrow, precisely at the point whence originate the principal nerves of the respiratory apparatus. It was with the view of causing this dislocation of the neck, and, consequently, of abridging the sufferings of criminals condemned to perish on the gallows, that hangmen were formerly accustomed to rest one foot upon the shoulder of the criminal at the moment he was thrust from the ladder with the cord around his neck. And from the same

33

cause, sudden death may happen in those imprudent plays where a child is raised from the ground by the ears.

The vertebral column, as we have already said, supports in a measure all the other parts of the body. By its superior extremity it articulates with the head, each of the dorsal vertebræ articulates with a pair of ribs, and the sacrum is wedged between the two haunch bones.

Head.   The head is composed of two principal portions, the cranium and face.

Cranium.   The cranium is a kind of osseous box, oval in form, occupying the whole superior and posterior part of the head, and lodging as before stated, the brain and cerebellum. Eight bones are united to form the walls, namely, the frontal or coronal in front (*f*), the two parietal (*p*), above, the two temporal (*t*), upon the sides, the occipital (*o*), behind, and the sphenoid (*s*), and ethmoid below. All these bones, with the exception of the last, have the form of large thin plates and are of a very compact texture, and all articulated together so as to be completely immovable and to give the cranium great solidity. These articulations are very remarkable in that they vary in form in the different parts of the cranium, for the purpose of better resisting external violence which might tend to disunite these bones, and would produce different

Fig. 41.[1]

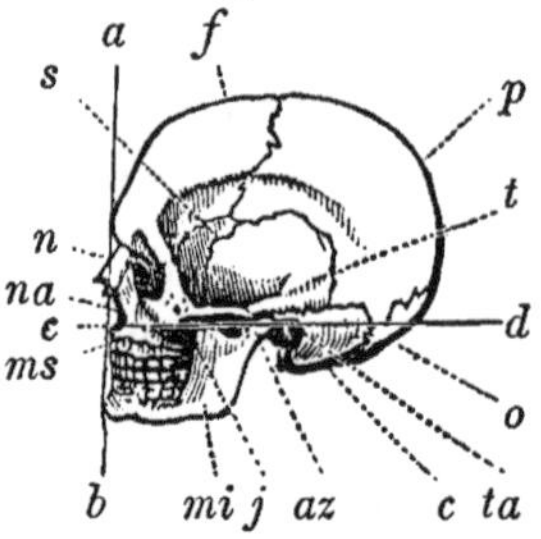

---

[1] *f*, Frontal bone, — *p*, parietal, — *t*, temporal, — *o*, occipital, — *s*, sphenoid, — *n*, nasal, — *m s*, superior maxillary, — *j*, malar or cheek bone, — *m i*, inferior maxillary, — *n a*, anterior opening of the nasal fossæ, — *t a*, auditory foramen, — *a z*, zygomatic arch formed by a part of the temporal and malar bones, — *a*, *b*, *c*, *d*, lines indicating the facial angle.

effects, according to the point on which it acts. Thus when a blow falls upon the summit of the head, the movement is propagated in all directions, and tends to separate the parietal bones by driving forwards or backwards the frontal and occipital bones; therefore, all these bones are united by closely serrated sutures. But when the cranium receives a blow upon the side, the effort acting upon the temporal tends to drive in this bone, to prevent which accident, nature has united the temporal to the neighboring bones, not by means of sutures, merely adapted to prevent disunion, but by a very oblique articulating border, so as to render this bone externally much larger than the space in which it is set.

The vault of the cranium presents nothing remarkable; but, at its base, we find a number of holes which serve for the passage of the blood vessels of the brain and of the nerves, which arise from the encephalon. Through one of these holes, hollowed in the occipital bone and much larger than the others, passes the spinal marrow, and there exists near its border upon each side, a large convex process called the *condyle*, which serves for the articulation of the head upon the vertebral column. The head is nearly in equilibrio upon this kind of pivot, but yet the portion in front of the articulation is more voluminous than that behind, and tends to counterbalance the first. Therefore, the muscles which extend from the vertebral column to the posterior part of the head, and which serve to raise the latter, are much more numerous and powerful than the flexor muscles, p'aced in the same manner in front of the column; and when the former are relaxed, as happens during sleep, the head usually leans forward and rests upon the chest.

Upon the sides of the base of the cranium, we also observe two very large processes, called *mastoid*, into which are inserted two muscles, which descend obliquely to the chest at the anterior part of the neck, and serve to turn the head upon the vertebral column.[1] Finally, directly in front of these processes, is the opening of the external auditory duct, which, together with the several parts of the middle and internal ear, is hollowed into that portion of the temporal bone called, from its hardness, the *petrous*.

Face. The *face* is formed by the union of fourteen bones of very different forms, and presents five great cavities for lodging the organs of sight, smell and taste. All these bones, except the lower jaw, are completely immovable, and are articulated together or with the bones of the cranium. The two principal are the *superior maxillary bones (ms)*, which constitute almost the whole of the upper jaw, and which articulate with the frontal so as to aid in the formation of the orbits and nasal fossæ. On the outer side they are articulated with the *malar bones (j)*, and behind with the *palate bones*, which in their turn are joined to the sphenoid.

The *orbits*, as we have elsewhere seen, are two conical fossæ or pits, the base of which is directed forward : the arch of these cavities is formed by a portion of the frontal bone and their floor by the superior maxillary bones. On the inner side, the ethmoid and a small bone, called the *lachrymal*, complete their walls, and on the outer, they are formed by the malar and the sphenoid ; the latter also occupies the bottom of the orbit, and in it are found

---

[1] Called, from their attachment, *sterno-mastoid muscles*.

the openings for the passage of the optic nerve, and the other nervous branches of the apparatus of vision. In the vault of the orbit may be seen a depression which lodges the lachrymal gland, and on its exterior wall is found a canal which descends vertically into the nasal fossæ and gives passage to the tears.

The *nose* is formed in a great measure of cartilage; thus in the skeleton the anterior opening of the nasal fossæ (*na*) is very large, and the bony portion formed by two small bones, called *nasal* (*n*), is but slightly prominent. The nasal fossæ are very extensive; superiorly, they are excavated in the ethmoid bone, the whole interior of which is filled with cellules; inferiorly, they are separated from the mouth by the arch of the palate, which is formed by the superior maxillary and two palate bones; finally, they are separated at the median line by a vertical septum formed superiorly by a plate of the ethmoid, and inferiorly by a particular bone called the *vomer.* In the interior of these fossæ may, also, be found two distinct bones which are called the *inferior turbinated* and also the opening of the frontal, sphenoidal, and maxillary sinuses, cavities more or less extensive, hollowed in the bones from which they take their names.

All the teeth of the upper jaw are implanted into the superior maxillary bone. In early life it is formed of several pieces, and in most animals we find an anterior portion which is called the intermaxillary bone.

The lower jaw, in man, is composed of a single bone, for the two halves of which it is formed in many animals are early united and completely consolidated. This bone, called *inferior maxillary*, much resembles a horseshoe, the curved ends of which are greatly raised. It is

articulated with the temporal bones by a projecting condyle situated on each extremity, and received in a cavity
called the *glenoid* (*e*); finally in front of these condyles
there rises upon either side a process called coronoïd,
which serves for the insertion of one of the levator muscles of the lower jaw (the temporal muscle); these muscles (*t*) are all fixed to the angle of the jaw, and at a
short distance from the fulcrum point on which this lever
moves. In the majority of instances, on the contrary,
the resistance to be overcome by this same lever during
mastication, is applied to the anterior part of the jaws;
therefore, these muscles, although very powerful, can produce but very feeble effects, and to crush between the
teeth the hardest bodies, they must be carried as far as

Fig. 22.

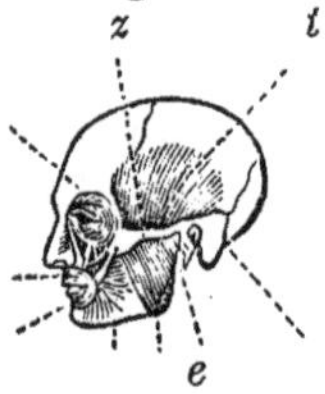

possible to the bottom of the mouth, so
as to shorten the resistance arm of the
lever and render it equal or even shorter
than that of the power. These muscles
are fixed on the internal, as well as external face, of the jaw, and make their
fulcrum point upon the sides of the head, as high as the
temples, passing between the lateral walls of the cranium
and a bony arch called the zygomatic (*z*), which extends
from the malar bone to the ear, and also serves for the
insertion of these organs.

The head, as before shown, is essentially composed of
twenty-two bones; but their number is actually greater;
for in the interior of each temporal bone, there are, as
elsewhere stated, four little bones belonging to the apparatus of hearing, and the *hyoid bone* may also be considered a dependence of the head. This bone is suspended
from the temporal bones by ligaments and is placed across

the upper part of the neck, where it bears the tongue and sustains the larynx.

The cervical vertebræ only articulate with each _{Thorax.} other, or with the head and the first dorsal vertebra ; but each of the twelve dorsal vertebræ supports a pair of very long and flat arches which curve around the trunk, so as to form a kind of osseous cage to lodge the heart and lungs. These arches are the ribs, the num- _{Ribs.} ber of which is consequently twelve on each side of the body; their posterior extremity is articulated with the body of the corresponding vertebra and with one of the transverse processes ; the other extremity is continuous with a cartilaginous shaft, which, in certain animals (birds, for example,) is always ossified, and is then called the *sternal rib.* The cartilages of the seven first pairs of ribs, which are called the true ribs, are united to the sternum, which is a single bone, occupying, in front, the median line of the body and serving to complete the walls of the thoracic cavity ; the five last pairs of ribs do not reach the sternum, but are joined to the cartilages of the preceding ribs, these are called the false ribs, (fig. 17).

Upon the bony cage just mentioned are fixed _{Superior limbs.} the *superior limbs.* In each of them may be distinguished a *basilar* portion, which may be compared to a pedestal upon which is inserted the essentially movable portion of the limb, which itself represents a lever, to which the former serves as a fulcrum. This basilar portion is composed of two bones, the scapula and clavicle.

The *scapula* is a large flat bone, occupying the _{Scapula.} superior and external part of the back : its form is nearly triangular, and it presents, on the upper and outer side, an articulating cavity quite large, but rather shallow, which

receives the extremity of the bone of the arm (glenoidal
fossa of the scapula). At its superior border we find a
projecting process called the *coracoid*, and on its external
face a very prominent horizontal crest, which terminates
above the articulation of the shoulder by a process called
the *acromion*, at the extremity of which the *clavicle* is
Clavicle. articulated. This latter bone is small and cylindrical; it is placed across the superior part of the chest
and extends, like an arch, from the sternum to the scapula. Its principal use is to keep the shoulders separated:
and, therefore, it is very often broken, when, by falls upon
the side, this part is driven with violence toward the sternum; and in those animals which bring the arm powerfully to the breast (as birds in flight), this bone is greatly
developed, while it is completely wanting in those who do
not execute such motions, and only move their limbs lon-
gitudinally, as in horses, etc.

Muscles of the shoulder. Numerous muscles fix the scapula to the ribs.
One of the principal is the *serratus magnus*, which extends
from the anterior part of the thorax to the posterior border
of this bone, passing between it and the ribs. In man it
is not much developed; but in quadrupeds it is extremely
strong and constitutes with that of the opposite side a
kind of girth, which supports the entire weight of the trunk,
and which prevents the scapulæ from ascending toward
the vertebral column. In man, the *trapezius muscle*, which
extends from the cervical portion of the vertebral column
to the scapula, has, likewise, very important functions: for
it raises the shoulder and sustains the weight of the entire
thoracic extremity, therefore it is greatly developed.

The portion of the thoracic extremity which constitutes
the lever to which the scapula serves as a fulcrum, is
composed of the arm, fore-arm, and hand.

The arm is formed by a single long cylindrical Humerus. bone called the *humerus.* Its superior extremity (or *head*) is large, round, and articulated with the glenoid cavity of the scapula, in which it can roll in any direction. The muscles which move this bone are inserted into its superior third and attached by their opposite extremity to the scapula or thorax. The three principal are the great pectoral, which carries the arm inward at the same time it depresses it ; the great dorsal, which carries it backward and downward, and the deltoid, which raises it.

The inferior extremity of the humerus is enlarged and has the form of a pulley, upon which the fore-arm moves as on a hinge.

Two long bones, placed parallel, form this por- Ulna and radius. tion of the thoracic extremity ; the *ulna* situated on the inner, and the *radius* on the outer side. They are united by ligaments and an aponeurotic septum, which extends from one to the other, their whole length : but yet they are movable, and the radius, which bears at its extremity the hand, can turn upon the ulna, which serves for its support. From the different uses of these two bones it might be foreseen what would be the principal differences in their general form. The *ulna,* to articulate in a solid manner with the humerus, must present at its superior extremity a certain size, and an extensive articulating surface, while at its inferior extremity, where it serves as a pivot to the radius, it must be small and rounded. The radius on the contrary must be, for the same reason, small at its superior, and very large at its inferior extremity, to which the hand is suspended. This is actually the case, and we also observe that these two bones only touch at

34

their two extremities, which renders more easy the motions of rotation of the radius upon the ulna.

The ulna which carries with it the radius, moves upon the radius in only one direction; it only executes motions of flexion and extension, and in the latter it can only form, with the humerus, a straight line, for it presents beyond its articulating surface a process, called the *olecranon*, which in that situation rests upon the humerus and affords an invincible obstacle to any farther extension. The extensor and flexor muscles of the fore-arm extend from the shoulder, or superior part of the humerus, to the superior part of the ulna: wherefore they are arranged in a manner favorable to the rapidity of motion of the fore-arm, but very unfavorable to the employment of a great force ; for the power-arm of the lever, represented by the space comprised between the articulation of the elbow and their insertion, is very short, while the resistance-arm of the lever, which is equivalent to the entire length of the limb from this same articulation, is, on the contrary, very great.

The rotation of the radius and of the hand upon the ulna are effected by muscles situated on the fore-arm, and which extend obliquely from the extremity of the humerus, or ulna, to one or other of these parts.

Hand.   The hand is divided into three portions, carpus, metacarpus, and fingers.

Carpus.   The *carpus* or wrist is formed by two ranges of little short bones, united very closely together, so that the whole of this part enjoys a degree of mobility ; although each of the bones composing it can scarcely be displaced, an arrangement which is adapted to give their articulations a very great solidity.   They are in number eight.

Four of these bones, viz.: the *scaphoid, semi-lunar, pyri-
midal* and *pisiform* compose the first range; the four
others, which are called *trapezium, trapezoid, os magnum,*
and *os unciforme,* form the second. It must be remarked
that these different bones are arranged so as to protect
the vessels and nerves passing from the fore-arm to the
hand; for this purpose they form, with the ligaments,
a canal which is traversed by these organs and which
can support, without flattening, the strongest pressure.

The *metacarpus* is composed of a range of little Metacarpus.
long bones, placed parallel-wise, and equal in number to the
fingers with which their extremities articulate. Four of
these bones are united by their two ends, and are scarcely
movable; but the fifth, which bears the thumb, only articulates with the carpus and moves freely upon the latter.

Finally, the fingers are formed each by a series Phalanges.
of small, slender bones, joined end to end, and called
phalanges. The thumb presents only two, but the fingers
have three. The last phalanx bears the nail. The fingers
are all very movable and may all move independently of
each other. Their flexor and extensor muscles form the
greater part of the fleshy mass of the fore-arm, and terminate by extremely long and small tendons, some of which
are fixed to the first phalanges, and others to the last.

When we examine the united thoracic extremities, we remark that the several levers joined end to end to form them,
diminish progressively in length. The arm, for instance,
is longer than the fore-arm; the latter is longer than the
wrist, and each phalanx is shorter than the preceding.
The utility of this arrangement is readily comprehended.
The numerous and close articulations at the extremity of ·
the limb, permit the latter to vary its form in a thousand

ways, and to accommodate itself to the body to be seized; while the long levers, formed by the arm and fore-arm, permit us to carry the hand rapidly to considerable distances. The motions of the humerus upon the scapula principally determine the general direction of the limb; the articulation of the elbow is specially designed to permit it to be lengthened or shortened.

Inferior members. The structure of the inferior members is very analogous to that of the thoracic, and the principal differences are those necessary for their greater solidity at the expense of their mobility, and to render them organs of locomotion rather than of prehension. They also have a basilar portion, which is the counterpart of the shoulder, and is called the *haunch,* and an articulated lever formed of three parts, the thigh, the leg and the foot, which answer to the arm, fore-arm and hand.

Iliac bone. The haunch, or basilar portion of the abdominal extremity, is formed by a large flat bone, called the *iliac* (from the latin word *ilia,* flank) or *coxal bone* (from the word *coxa,* which in Greek signifies haunch). This bone results from the union of three principal pieces, always distinct in fœtal life, which may be compared to the body of the scapula, its coracoid process and the clavicle. The iliac bones have not bases corresponding to the bones of the shoulder, ribs and sternum, to rest upon; being destined to sustain the entire weight of the body they must, however, be fixed in the most solid manner to the trunk; thus they are articulated behind with the portion of the vertebral column, called the sacrum, and in front they unite and form an arch, called the *pubis.* They are perfectly immovable, and from the union of these two bones and the sacrum results a large osseous cincture, which inferiorly terminates the abdomen, and which, from its open form, is called *pelvis* (fig. 52). This kind of ring is closed inferiorly by muscles and gives passage to the intestine called rectum, and to the genito-urinary organs. Upon the sides, externally, we observe upon each iliac bone an articulating cavity, nearly hemispherical, which serves to lodge the head of the thigh bone. Finally, most of the muscles which move the thigh and leg, are inserted into the pelvis, and the muscles, which enclose, as we have elsewhere seen, the abdominal cavity, are fixed to it, and reach from it to the thorax.

The thigh, as the arm, is composed of a single Femur. bone, which is called the *femur*. Its superior extremity is inclined inwards, and its head, which is rounded, is separated from the body of the bone by a contraction, called the *neck of the femur*. At the lower end of this neck, and at the point where it joins the body of the bone at an open angle, are several large tuberosities which may be felt through the flesh, and which give insertion to the principal motor muscles of the thigh; finally, its inferior extremity is very large and presents two condyles laterally, compressed and rounded from before backward, which slide upon the articulating face of the principal bone of the leg and only permit the latter to be bent, or extended, while the femur itself may move upon the pelvis in any direction.

The difference of the leg and fore-arm is much Tibia, fibula, and patella. greater. Besides the *fibula* and *tibia*, which are the two principal bones of which this part of the limb is composed, as the fore-arm of the ulna and radius, we find in front of the knee a third bone called the *patella*, which may be regarded as analogous to the olecranon process of the ulna, and which principally serves to elongate from the knee the tendon of the extensor muscles of the leg, and to render its insertion more oblique, an arrangement, which as already shown, must increase the power of its action. The foot not being required to execute motions of rotation as the hand, but requiring, to support the whole weight of the body, that it should present great solidity in its articulation, the two bones of the leg do not move upon each other, and the one which articulates with the femur and which represents the ulna (the tibia), is also the one to the extremity of which the foot is attached. The fibula, which is small and situated on the external side of the tibia, only serves, so to speak, to maintain the foot in its natural position and to prevent it from turning inward. Its superior extremity is applied against the head of the tibia, and its inferior constitutes the outer ancle.

Foot.   The foot is composed, as well as the hand, of three principal parts, namely, the tarsus, metatarsus, and the toes.

Tarsus.   There are seven bones in the tarsus, and it only articulates with the leg by one of them, the *astragalus*, which ascends above the others and presents a head in form of a pulley, adapted to the cavity formed by the articulating surface of the tibia and the two malleoli.[1] The astragalus rests upon the *calcaneum* which is prolonged farther behind and constitutes the heel ; lastly a third bone, called *scaphoid*, terminates the first range of the bones of the tarsus, and the second range is composed like the hand, of four small bones, three of which

---

[1] The inner malleolus, or ancle, is a process of the tibia; the outer is formed by the fibula.

have received the name of *cuneiform bones*, and the fourth, placed in front, is called the *cuboid*.

The bones of the metatarsus, to the number Metatarsus. of five, exactly resemble those of the metacarpus: only they are stronger and less movable, especially the inner, which is arranged in the same manner as the rest. Phalanges. It is the same with the toes; we reckon the same number of phalanges as in the fingers of the hand, but they are shorter and not so easily moved. The great toe is not detached from the others and cannot be opposed to them, as the thumb to the other fingers.

Upon the internal side of the foot, the bones of the tarsus and metatarsus form a kind of arch to lodge and protect the nerves and vessels, which descend from the leg to the toes. When this arrangement does not exist, and the sole of the foot is flat, as sometimes happens, these nerves are compressed by the weight of the body, and walking cannot be long continued without pain. The whole extent of the foot, however, rests Muscles of the leg. upon the ground, and forms a large and solid base of support; it can be moved upon the leg only in the direction of its length, and the muscles serving for this purpose surround the tibia and fibula. The extensors of the foot, which form the calf of the leg, are fixed to the calcaneum by a long tendon, called the *tendon of Achilles*, and are arranged in the manner most favorable to their action; for their insertion is nearly at a right angle and farther removed from the fulcrum than the resistance to be overcome, when the weight of the body, pressing upon the astragalus, is raised by the foot.

All the mammiferæ, birds, reptiles, and fishes, Animals with an internal skeleton. have an internal skeleton, more or less nearly resembling that of man, composed of nearly the same bones, and likewise moved by muscles placed between this solid frame and the tegumentary envelope. This skeleton gives to their body its general form, and upon its disposition and the action of the muscles fixed to its various parts, depend the attitudes as well as motions of these animals.

Standing. A small number of these beings lie with the whole length of their bodies upon the ground, and only move by undulations of the trunk; but others are ordinarily sustained upon their limbs, and the name of *standing* is given to the state in which an animal thus supports itself on the ground by its legs.

That the limbs may remain firm and thus sustain the body, their extensor muscles must be kept in a state of contraction, for, without it, these organs would bend under the weight they support and thus occasion its fall. We have already seen that the muscles are more quickly fatigued the longer each contraction continues; thus in most animals, standing for a long time is more fatiguing than walking, in which latter case, the extensor and flexor muscles mutually relieve each other.

This condition is not the only one indispensable to standing; that the body may remain upright on the limbs thus stiffened, it must be in equilibrium.

An equilibrium is established, not only when a heavy body rests upon a resisting object with the whole extent of its largest surface; but also, when it is so placed, that if one part of the mass be depressed toward the earth, an opposite part, equally heavy, would be raised in the same degree; the weight of one part serves then to counterbalance that of the other, and the point around which all these parts are reciprocally in equilibrium and where support must be given to keep in place the entire mass, is called the *centre of gravity*. To support this centre, it is requisite that the *base of support* (that is to say, the space occupied by the points by which the mass leans upon a resisting object, or that comprised between these points), be situated vertically beneath it.

That the body of an animal may remain in equilibrium upon its paws, the vertical line passing through its centre of gravity must consequently fall within the limits of the space left between the feet, or covered by them : and the larger this base of support is in relation to the height at which the centre of gravity is found, the more stable will be the equilibrium, for the more it may be displaced without throwing the line of gravity, just mentioned, beyond the limits of this base. It is also deserving of note, that the more difficult the equilibrium to be pre-served, the more intense must be the muscular contraction to maintain it, and the more fatiguing the position of the animal.

From this reasoning it might be inferred, that when an animal rests upon all-fours, its standing will be more firm, solid, and less fatiguing than when it only rests upon two, and that in this latter case, its equilibrium will be more stable than when it only rests upon one leg : for the extent of the base of support is thus constantly narrowed. When an animal is supported upon four legs, the space comprised between them is very considerable, and can be but slightly modified by the larger or smaller extent of the surface of these organs. To render them very large would then have increased their weight, without adding to the solidity of the base of support:

35

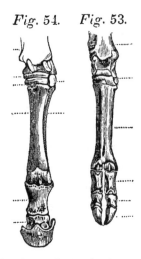

Fig. 54.   *Fig.* 53.

therefore, in most quadrupeds, the members touch the ground only by an extremity, hardly dilated; and we find the number of fingers constantly diminishing, without injury to these organs as instruments of locomotion: the foot of the stag and horse are cases in point (fig. 53 and 54); but when the animal only rests upon two of its feet, whatever be their distance apart, the base of support can have no solidity from before backwards, only in as much as these organs touch the ground in a considerable extent, as in the foot of man; and when an animal readily supports itself upon one paw, as birds, nature must have given their feet more width, as well as length.

To allow an animal to put itself in equilibrium upon a single leg, the foot on which it rests must also be placed vertically below the centre of gravity of its body, and its muscles so arranged as to permit it to keep this limb inflexible and immovable. Man can do this; for the centre of gravity of his body is about the centre of his pelvis, and when placed in a vertical position it is enough for him to lean a little to the side not supporting the weight, in order that the line of gravity may fall upon the sole of the foot of the opposite side; but with most quadrupeds the thing is impossible.

Most of these latter animals cannot even remain erect upon their hind paws, on account of the direction of the members relatively to the trunk; and if they do succeed for an instant, it is impossible for them to remain so, because their base of support is very narrow, the centre of gravity of their body is placed very high (toward the chest), and the muscles called into action in this attitude must be contracted with a violence requiring immediate repose In man and a small number of other mammiferæ, the vertical position upon the two abdominal limbs, is, on the other hand, more or less easy; for these members may readily be placed in the direction of the axis of the body, the centre of gravity is situated very low, and the base of support formed by the feet is very large. In man especially, is this attitude rendered solid by the width of the pelvis, the form of the feet, and several other corresresponding peculiarities of organization.

In the vertical position, the muscles of the posterior part of the neck are contracted to keep the head in equilibrium upon the vertebral column, the extensor muscles of which are also called into action, to prevent it from yielding under the weight of the thoracic members and of the viscera of the trunk, which tend to curve it forward. The whole weight of the body is thus transmitted by the vertebral column to the pelvis, and from the pelvis to the femur. Left to themselves, these latter bones would bend upon the pelvis, and the trunk would fall forward, did not the contraction of their extensor muscles keep them extended. The extensor muscles of the leg, at the same time, prevent the knees from bending, and the extensors of the foot keep the leg in a vertical position, so that the weight of the body is transmitted from the thigh to the leg, from the leg to the foot and from the foot to the ground.

The seated position is less fatiguing than standing, be-

cause the weight of the body being then directly transmitted from the pelvis to the base of support, it is not
necessary for the extensors of the abdominal extremities
to contract in order to maintain the equilibrium. Finally,
when man is laid upon his back or belly, the weight of
each movable portion of his body is directly transmitted
to the ground, and consequently, to continue so, he is not
required to contract a single muscle.

Locomotion. The progressive motions by which man and
animals are transported from one place to another, require
a determinate rapidity in a certain direction to be impressed upon the centre of gravity of their bodies. This
impulse is given to it by the employment of a certain
number of articulations, more or less bent, and the position of which is such that upon the side of the centre of
gravity their employment is free, while in the opposite
direction it is confined, or even impossible, so that the
totality or greatest part of the motion produced, takes
place in the first of these directions. The same thing
then takes place, as in a spring with two branches, one
extremity of which is applied against a resisting obstacle,
and the two branches of which, after being approximated
by a force from without, are restored to their former lib-
erty; by reason of their elasticity they would tend to
separate equally, until restored to their position before
they were approximated; but that applied against the
obstacle, not being able to overcome it, the movement
will take place in an entirely opposite direction, and the
centre of gravity of the spring will be repelled from this
obstacle with a greater or less degree of rapidity. In the
body of animals, the flexor muscles of the part employed
in each kind of motion, represent the force which com-

presses the spring, the extensors represent the elasticity which tends to separate the branches, and the resistance of the ground, or fluid in which these beings move, represents the obstacle which opposes the extension of one of the branches.

Walking is a motion upon a fixed plane, in Progression. which the centre of gravity is alternately moved by one part of the locomotory organs, and sustained by the others, without entirely preventing at any time the body from resting upon the ground. This latter circumstance distinguishes it from leaping and running, motions in which the body quits momentarily the ground, and throws itself into the air.

In walking upon two feet, in man and the other animals to whom this mode of locomotion is possible, one of the feet is carried in front, while the other is extended upon the leg ; and as this latter leans upon a resisting plane, its elongation displaces the pelvis and projects the whole body forward ; the pelvis turns at the same time upon the femur of the opposite side which sustains it, and the leg, which previously remained behind, is bent and carried in front of the other, then straightens, and in its turn serves to sustain the body while the other limb, during its extension, gives a new impulse to the centre of gravity. By these alternate motions of extension and flexion, each limb bears, in its turn, the weight of the body, as in standing upon one foot, and at every step the centre of gravity is pushed forward ; but we see that it must, at the same time, be carried alternately to the right and to the left, to be directly above each of its bases of support.

The majority of quadrupeds in walking make use principally of their hind legs to propel their bodies forward,

and of their anterior legs to sustain themselves in the new
position given by each step. When these motions are
made by all four feet at the same time, the animal is for
an instant suspended above the ground, and this mode of
locomotion constitutes the gallop. In walking, two feet
only contribute to the formation of each step, one in front,
the other behind; in general, those upon opposite sides
are raised simultaneously, at others, those on the same
side; this latter gait is the amble.

Leaping is performed by the sudden employment of
various articulations of the limbs, which are at first more
flexed than usual. The extent of space the animal thus
passes through in the air, depends principally upon the
rapidity impressed upon its body at the moment of departure, and this rapidity, in its turn, depends upon the
proportional length of the bones of these members and
the force of their muscles. Therefore, animals which leap
the best, have the hind legs and thighs very long and
very muscular.

Swimming and flying. Swimming and flying are motions analogous to
those of leaping; save that they are effected in fluids, the
resistance of which takes the place, to a certain degree,
of that of the ground in the preceding phenomena.

The members, which by extending and bending backwards drive the body forward, are applied in this case
against the water or air, and tend to crowd together the
particles of which these media are composed with a
greater or less rapidity; but if the resistance, which the
air or water presents in this way, is superior to that which
opposes the motion of the animal in a contrary direction,
these fluids will furnish to the limb a fixed point, and the
motion produced will be the same as if this spring, touch-

ing by its posterior extremity an invincible obstacle, only expanded with a force equal to the difference existing between the rapidity employed and that impressed by it upon the surrounding fluid in bending backwards. Now, the less dense the fluid in which the animal moves, the less resisting will be the fixed point furnished by it ; and the greater the force necessary to overcome, in rapidity, this displacement of the fixed point, and propel the body forward. Thus, flying requires a much greater motor power than swimming, and neither of these movements could be effected by the force requisite for a leap upon a solid surface. But this great employment of the motor force is not the only condition necessary for aërial or aquatic locomotion ; as the animal plunged in a fluid, finds upon all sides an equal resistance, the rapidity acquired by striking upon this fluid behind, would soon be lost by that which it would be obliged to displace in front, if it could not diminish considerably the surface of the locomotory organs immediately after having made use of them to give the blow. This is the actual case, and one of the characteristics of every organ of flight, or even of swimming, is the power of changing its form and presenting, in the direction perpendicular to that of the movement produced, a surface alternately very large and very narrow.

With regard to the structure of the organs of aërial, or aquatic locomotion, we will only say that in the superior animals the thoracic almost always alone serve for flight, and that their transformation into wings takes place either by the extreme elongation of the fingers, and the existence of a membrane which extends between these appendages, and is also fastened to the sides (as in the bats, fig. 55), or by the implantation of long and stiff feathers upon the whole extent of the limb, which then becomes

Fig. 55.

Fig. 56.

long and narrow (as in birds). The abdominal and thoracic members may also serve for swimming, and when they are completely transformed into fins we find, in general, that their terminal part becomes very large, and that the portion which represents the arm and forearm is shortened, so that the portion analogous to the hand seems to spring directly from the trunk: this is easily proved in the phocæ and cetacea (fig. 56).

THE VOICE.

In concluding the history of the functions of relation, it it remains for us to treat of the production of sounds, a faculty which in man is of extreme importance, as upon it depends voice and speech.

In the lower animals there is no trace of this faculty, and in *insects*, the monotonous noise which is called the hum of these little beings, merely results from the rubbing of their wings, or some other parts of their tegumentary envelope, against each other; but the superior animals nearly all can utter sounds more or less varied, the production of which depends upon the passage of the air into a determinate part of the respiratory duct, so arranged as to impart vibration to this fluid.

Larynx.   In man and the other mammiferæ, this phenomenon takes place in that portion of the air passage which is situated at the top of the neck, between the pharynx and the trachea, and called the *larynx* (fig. 16,

a; fig. 23, e; fig. 24, g). In fact an opening made in the trachea below this organ, by permitting the expired air to escape outwardly without traversing it, completely prevents the production of sounds. Examples are cited of persons who with such an opening in the neck, produced either by a wound or some disease, have lost the voice, but recovered the faculty of speech by putting a close cravat around the neck so as to stop the wound, and, consequently, force the air to follow its usual course. On the other hand, a similar opening above the larynx does not destroy the voice; whence we may with certainty conclude, that the production of sounds takes place in this organ.

The *larynx* is a large and short tube, which is suspended to the os hyoides, (h) and continues inferiorly into the trachea. Its walls are formed by various cartilaginous plates, designated as the *thyroid cartilage* (t), the *cricoid cartilage* (c), and the *arytenoid cartilages*; in front we see the projection usually known as Adam's apple (a); and in the interior, the mucous membrane lining it, forms, about its centre, two great lateral folds, directed from before backwards and arranged like the lips of a button-hole. These folds are called *vocal cords*, or *inferior ligaments of the glottis*; they are very thick;

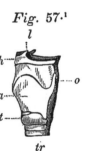

Fig. 57.[1]

[1] Profile view of the human larynx; h, os hyoides;—l, body of the hyoid bone, to which is attached the base of the tongue;—t, thyroid cartilage;—a, projection formed by the thyroid cartilage in front, and commonly called *Adam's apple;* the thyroid cartilage is united to the os hyoides by a membrane;—c, cricoid cartilage;—tr, trachea;—o, posterior wall of the larynx in relation with the æsophagus.

Fig. 58.[1]

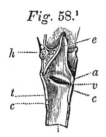

their length is in proportion with the projection of the anterior part of the thyroid cartilage (or Adam's apple), and by the aid of the contractions of a small muscle lodged in their interior, and of the movements of the arytenoid cartilages to which they are fixed posteriorly, they may be approximated in a greater or less degree, so as to enlarge or diminish the space of the fissure (the opening of the glottis), which separates them. A little above the vocal cords, are found two other similar folds of the mucous membrane of the larynx; they are called *superior ligaments of the glottis,* and the two depressions which divide them from the inferior ligaments, are named *ventricles of the larynx.* The space comprised between these four folds is called the *glottis;* finally above this opening is a kind of fibro-cartilaginous tongue, the *epiglottis,* fixed by its base beneath the root of the tongue, and which ascends

Fig. 59.[2]

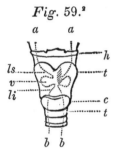

obliquely into the pharynx, but which can be depressed so as to cover the glottis, as was exemplified when treating of deglutition, (fig. 58, *e*).

In the ordinary state, the air expelled from the lungs

[1] Vertical section of the larynx; — *h,* os hyoides; — *t,* thyroid cartilage; — *c, c,* cricoid cartilage; *a,* arytenoid cartilage; — *v,* ventricle of the glottis, formed by the space left between the vocal cords and the superior ligaments of the glottis —; *e,* epiglottis.

[2] Front view of the larynx; the shape of the internal wall is indicated by the lines *a, a, b, b;* — *li,* inferior ligaments of the glottis or vocal cords; — *ls,* superior ligaments. The other parts are indicated by the same letters as in the preceding figures.

traverses freely the larynx and produces no sound in it; but when the muscles of this organ contract and the passage of the air becomes more rapid, the voice may be heard. An experiment by Galen demonstrates the necessity of these contractions for the formation of sounds.

He divided in living animals the nerves supplying the muscles of the larynx;[1] and this operation, which occasioned paralysis of these organs, also involved the loss of the voice. Other experiments also prove that the production of sounds depends especially upon the action of the ligaments of the glottis. When the superior folds are divided, the voice is considerably weakened, and when the inferior, or vocal cords, it is destroyed.

Most physiologists consider the larynx as acting in the production of the voice upon the principle of a reed instrument: they think, that the air expelled from the lungs rushes out, in a continual stream, from the lips of the glottis, until these elastic cords return upon themselves and momentarily close the respiratory passage, to be separated anew, so as to produce movements of vibration sufficiently rapid to give birth to sounds, nearly the same things that transpire in blowing upon the reed of a hautboy. But from the recent observations of M. Savart it would appear, that the production of the vocal sounds does not depend upon a mechanism similar to that of the reed instrument, but rather takes

Mechanism of the production of sounds.

---

[1] The pneumogastric nerves, which arise from the lateral portion of the medulla oblongata, issue from the cranium, descending one upon each side of the neck, and penetrate into the thorax and abdomen. Immediately after their entrance into the former of these cavities they give origin to a branch, which ascends on either side of the neck and ramifies upon the larynx; it is called the recurrent nerve, from its direction.

place in the same manner as in the little instruments used
by sportsmen to imitate the song of birds.

These instruments, called *bird-calls*, are ordinarily
constructed of wood or metal, and consist of a small
cylindrical canal, very short and closed at each end by a
thin plate, pierced in its centre by a hole. To elicit
sounds, the sportsman places the call between his teeth
and aspires the air through the two openings by which it
is pierced. The current, which thus traverses the instrument, draws with it a part of the air contained in its
cavity, and the latter, being rarefied, soon ceases to be in
equilibrium with the pressure of the atmosphere, which
by its reaction crowds upon and compresses it, until, by
its own elasticity and by the influence of the current, it
undergoes a new rarefaction, followed by a second con-
densation, and so on. The small mass of air contained
in the call thus enters into vibration, and gives origin to
sonorous waves, which scatter themselves in the atmosphere. By moderating, or accelerating the rapidity of
the current, we produce grave or acute sounds, and they
are varied yet more by enlarging or contracting the openings of the instrument, varying its form, rendering its
walls more or less elastic, and adapting to it tubes of varying length.

It would appear that, by similar modifications of the
larynx, the sounds produced by this organ become grave
or acute. As the voice ascends, the lips of the glottis
become tense and contract so as to diminish, to a great
extent, the size of the opening left between them.
The contraction of the muscular fibres upon the walls of
the ventricles of the larynx, and of the muscles of the
posterior fauces, give at the same time to all these parts

a degree of tension favorable to the development of the sound produced, and we find that the larynx itself ascends as the sounds are more acute, a circumstance, which is explained by the laws of acoustics, for it causes the shortening of the passage traversed by the sounds in making their exit, and we know perfectly well that in our ordinary instruments of music the length of this passage has the greatest influence upon the rapidity of the sonorous vibrations; when we wish to draw from the reed of a clarinet, or hautboy, for example, a succession of sounds, we lengthen or shorten the tube formed by the body of the instrument, by closing or opening the holes with which the walls are pierced.

The intensity or volume of the voice depends in part upon the force with which the air is expelled from the lungs, in part upon the facility with which different parts of the larynx vibrate, and upon the extent of the cavity in which the sounds are produced.

The same individual cannot make heard, with an equal degree of force and clearness, all the sounds his larynx is capable of producing, because the different parts of his vocal apparatus are not arranged in a manner equally favorable to their production. When a man is weakened by fatigue or illness, his voice loses its intensity, because the muscles which drive the air from the lungs can no longer expel it with their ordinary force.

Finally, it is to the more considerable volume of the larynx in man, that a part of the remarkable difference between his voice and that of woman must be attributed; and owing to the existence of great cavities in communication with this organ, the howling apes, and some other animals, are able to make their deafening cries heard to an immense distance.

The pitch of the voice appears to belong in part to the physical properties of the ligaments of the glottis and walls of the larynx, and in part to that of the following portion of the vocal canal. We know from experiment that the pitch of musical instruments varies greatly, according as they are constructed of wood, metal, etc., and a coincidence has been remarked between certain modifications of the human voice, and the more or less hardened state of the cartilages of the larynx. In women and children, whose voices have a peculiar pitch, the cartilages of the larynx are flexible and have but little hardness, while in men and women with a masculine voice, the thyroid cartilage is remarkable for its power, and more or less complete ossification.

The form of the external opening of the vocal apparatus exerts a great influence upon the tone of the sounds produced. When the sounds traverse the nasal fossæ only, they are disagreeable and nasal ; when the mouth is wide open the voice acquires, on the contrary, power and brilliancy, and it would appear that the degree of tension of the velum palati and of the other parts of the fauces, exercises an influence not less great upon the manner in which sounds are modulated.

From what has been said of the mechanism of the production of sounds, it must be inferred that the diapason of the voice depends in a great degree upon the length and thickness of the vocal cords. The voice of man, as every one is aware, is much graver than that of woman ; because, in man, where the larynx makes at the superior part of the neck a considerable prominence, known by the vulgar name of Adam's apple, these folds are much longer than in woman, where the antero-posterior diameter of this organ is so small that the eminence alluded to can hardly be distinguished.

The sounds produced by the vocal apparatus have not always the same character, and are consequently distinguished into cries, singing, and the ordinary voice.

The *cry* is a sound usually acute and disagreeable, Cry. which is but little or not at all modulated, and which principally differs from the other vocal sounds by its pitch. It is the only one formed by the majority of animals, and, in this respect, man differs from the latter only by education. The infant just born can utter only cries, and when deprived of the sense of hearing its voice does not change; but when it understands what passes around it, it learns from its equals to modulate it, and to produce sounds of a particular nature.

This *acquired voice* differs from the cry by its Acquired voice. intensity and pitch; but it is not formed in the same manner as sounds, the intervals and harmonic relations of which are not clearly distinguished by the ear. Singing. *Singing*, on the contrary, is composed of appreciable or musical sounds, in which, so to speak, the ear reckons the relative number of vibrations.

Man also possesses the faculty of modifying, in a particular manner, the various sounds of his voice; he may articulate these sounds, and to this act is given the name of *pronunciation*.

The organs of pronunciation are the pharynx, Articulation of sounds. nasal fossæ, and the different parts of the mouth; and according to the manner in which they act, the sound produced by the larynx varies in its character, and constitutes a particular articulated sound.

The articulated sounds are divided into two great

classes, vowels and consonants. The former are permanent and simple sounds, which cannot be confounded by
union with others, and during the production of which
the apparatus of pronunciation remains unchanged ; the
consonants are, on the contrary, articulated sounds
which it is impossible to prolong as the vowels, and
which require for their production, particular movements
of the apparatus of pronunciation, movements in the
course of which this apparatus necessarily takes the disposition by means of which it forms a vowel. Thus, the
consonants can only be articulated by the addition of the
sound of a vowel. They are divided into labial, dental,
nasal, etc., consonants, depending upon the part called
into action in the mechanism of their pronunciation,
whether the lips, tongue, etc.

Man is not the only animal having the faculty of
articulating sounds, and thus to pronounce words ; but
he is the only one which attaches a meaning to the words
pronounced and to the arrangement he gives them ; he
alone is endowed with the power of speech.

FUNCTIONS OF REPRODUCTION.

The various vital phenomena, which have hitherto
occupied our attention, all have respect to the preservation of the life of the animal, or to its relation with surrounding objects. Those now remaining to be noticed
are of another order ; their object is the multiplication of
the individuals and the preservation of the species. We

have as yet only studied beings completely formed ; we are now to inquire into their origin, and the mode of their development.

The primitive creation of organized beings is ^{Creation of animals.} a subject which at first sight appears inaccessible to science ; but the genius of a man, of whom France may justly boast, Cuvier,[1] has dispersed a part of the profound darkness which enshrouded this great mystery, and taught us, at least, the order in which the different animals successively appeared upon the surface of our globe.

This earth has not always had the configuration it now has. Its various parts have been, on several occasions, invaded and abandoned by the waters, and by each inundation solid matters have been deposited, which, in the course of time have formed more or less thick layers of stone, clay, sand, etc., placed one above the other, according to the recent period of their formation. In the primary strata or formations, no trace is found of the remains of organized beings ; but as we ascend and explore the layers of more recent formation, we meet in the fossil state with wood, leaves, shells and bones of various forms ; sometimes these remains are in such quantity that the stone seems formed by them, and their preservation is often so perfect that it is easy to determine to what plants, or animals, they must belong.

The study of these remains, which have survived all

---

[1] M. George Cuvier was born at Montpellier in 1769, and died at Paris in 1832. His principal works are : Researches upon Fossil Bones, 5 vols. 4to; Lessons in Comparative Anatomy, 5 vols. 8vo; his general classification of animals entitled the Animal Kingdom, divided according to its organization, 5 vols. 8vo; and his Memoirs upon the Anatomy of the Mollusca, 1 vol. 4to ; but science has also been benefited in many ways by his indefatigable labors.

the great catastrophes so often overwhelming the surface
of the earth, demonstrates that our globe, originally destitute of living beings, was at first peopled by vegetables
only, and that the first animals which appeared were
beings of a very inferior structure, the species of which
have been for a long time destroyed.   They were probably aquatic animals, analogous to the sow-bugs and.
called by naturalists *trilobites*, or marine mollusca, living
in large shells, and called *ammonites*, or horns of Ammon.
At a more recent period, our globe has been inhabited
not merely by conchiferæ, analogous to those formerly
existing, but also by many enormous reptiles, frequently
of a most singular form.   The mammiferæ do not appear till some time after, and those which have successively peopled the earth have approached nearer and
nearer to the species actually inhabiting it.   Finally man
appears to have been the last, as he is the most perfect
of these creations, for we find none of his bones in the
fossil state.

Thus nature has gone on constantly complicating and
perfecting her works, and, by many intervening degrees,
she has ascended from the production of a plant to that
of a man.

At every great catastrophe of nature, several of the
existing species have been completely destroyed and replaced by new ones, so that each of the epochs of the
antediluvian history of our globe is characterized by its
particular population ; but other species appear to have
survived these changes ; and, according to the opinion of
certain naturalists, some of the animals which then appeared were of the species now existing, but modified in
their structure by the influence of the new condition in
which they are placed.

However the case may be with these entirely new creations, or with these transformations of beings already existing, it is evident, that from the most remote period of history, animals have succeeded each other with the same forms. At the present day we find in the catacombs of ancient Egypt, mummies of men, crocodiles, and several other animals, which have been buried in them for two or three thousand years, and which resemble in all points the individuals now existing.

Their mode of origin also explains in a degree this reproduction of the identical forms for a long succession of ages, while the beings which present them, all perish at the expiration of a short time, or even a few days. They actually at first form a part of the body of another organized being, which transmits life to them, and is in some sort the model after which they are constituted.

It was formerly a received opinion, that mat- Spontaneous generation. ter placed in favorable physical conditions might be organized of itself, become the seat of a vital movement, and thus give birth to animals but slightly complicated, such as the flies upon carcasses, etc.; but at the present day we know positively, that in the immense majority of cases, if not always, animals can only arise from parents similar to themselves; and that if spontaneous generation be possible, it is only in beings with the simplest structure, as the monads, which appear in the infusions of vegetable and animal matter, and which seem to be mere agglomerated organized globules, in a larger or smaller number, and endowed with motion. Their mode of formation is a disputed point among naturalists; according to some, these infusies, as well as many other microscopic globules, are exceptions to the general law, and are formed spontaneously ; while, according to others, they always arise from the germs of individuals similar to themselves, abundantly diffused in nearly all organic substances, and only developed under certain favorable circumstances.

Normal generation. But if it is the essence of organized beings to arise from parents like themselves, the manner in which this reproduction is effected varies greatly in the different animals, and we are thus furnished with a striking example of the application made by nature of the principle of the division of labor, when she has desired successively to perfect the beings of her creation.

Generation by slips. In the simplest animals, the important function of generation does not appear to be confided to any organ in particular ; all parts of the surface of the body of one of these animals are capable of giving origin to small slips, which increase in size, and soon become new individuals, similar in every respect to that from which they spring. The polypi, of which we have already had occasion to speak (fig. 1) are thus reproduced by slips ; but this mode of propagation of the species is only met with in a very small number of the most inferior animals,

Oviparous generation. and in all others the germ, which by its development is to constitute the young, is formed in a particular organ, called the *ovary*.

When these organs first display themselves in the zoölogical series, they have a very simple structure ; they are in general mere glandular vessels, and the germs they produce are adapted for development without the consent of any other apparatus ; but the division of labor is soon extended ; and reproduction is entrusted to two distinct organs, the consent of which is necessary to the birth of a new individual.

These two kinds of apparatus, one serving for the production, the other the fecundation of the germ, and called female and male apparatus, are at first united in the same individual, which of itself alone is charged with the whole labor of reproduction. Oysters, muscles, and many other inferior animals present this mode of generation. In others, where the sexes are still united, snails, for example, hemaphroditism is less complete : for the fecundation of the germs can only be effected by the individual producing them. But, as we ascend in the series of beings, we find nature pushing yet farther the division of labor ; for then the sexes are always separated. This is the case with all the superior animals, quadrupeds, birds, fishes, etc., and even with insects, spiders, the crustacea and some mollusca.

In most fishes, and even some reptiles, the fecundation of the germ does not take place till after the expulsion of the ova, and is in some degree confided to chance ; but in the superior animals, it is better ordered and takes place before their expulsion. In general, the germ, after being detached from the ovary and fecundated, has no longer need of the assistance of its parents for development. Abandoned to itself, it gives birth to a new individual, which carries with it the materials necessary for its nutrition during the entire life of the embryo ; but in birds, the egg, which is composed of the germ, nutritive substances just mentioned, and the membranes containing them, is only developed by the influence of ^{Viviparous generation.} an elevated temperature, which the mother ordinarily maintains by incubation. Finally, in other animals the series of the phenomena of reproduction is yet more complicated ; for the germ does not carry with it its nourishment, and to live after being detached from the ovary and fecundated, it must contract new vascular adherences with the walls of a particular sac, called the womb, which is destined to lodge the young individual until all its organs are formed.

The animals, whose germs do not thus draw their nutrition from the blood of the mother, are called *oviparous*; those which present the latter mode of reproduction are called *viviparous*, because in place of being developed in an egg they are born alive and ready formed.

All the inferior animals which are not reproduced by slips, such as worms, the mollusca, the crustacea, insects, etc., are oviparous: so are fishes, reptiles, and birds; but man and all animals closely allied to him, such as the domestic quadrupeds, etc., are viviparous.

Development of the fœtus. When the young individual begins to be developed in the germ, it is not, as might be supposed, the miniature of what it will become at a later period. It does not at all resemble its parents, and has neither its future form, nor structure. Its organs appear in succession, and they undergo during their evolution very remarkable changes. We may say in a general manner, that the entire organization of the embryo, as well as each of its parts considered separately, passes through a series of transitory stages, which resemble to a certain degree the permanent condition of some animal less elevated in the scale. The human embryo, for example, in the earliest moments of its existence presents a mere round mass, destitute of limbs, having some analogy in structure to certain very simple animals; for we find in it neither brain, heart, bone, nor distinct muscles. The

heart at first resembles that of some worms, a mere vessel, which soon curves and presents two dilatations, forming the left auricle and ventricle. It then exhibits a
mode of conformation analogous to that of some fishes;
the auricle is afterwards divided into two cavities by an
incomplete partition, which resembles the structure of
the heart in most reptiles, and at a little later period, a
second partition which arises from the bottom of the
ventricle also divides this into two, so that the heart then
presents two cavities as we find it in the superior animals. But yet the circulation of the fœtus is closely
allied to that of the reptiles; for the two auricles communicate by a hole called the *foramen ovale*, and the
pulmonary artery is joined to the aorta by a large anas-
tomotic branch, so that only a small portion of the blood
driven from the right ventricle reaches the lung, while
the remainder mixes with the blood destined for the immediate nutrition of the organs.

The development of the embryo can be more closely
watched in the egg, therefore we shall select it as an example for the study of this curious phenomenon.

Birds have not, as most of the superior animals, two
ovaries. We find in them only one, which is fixed in the
abdominal cavity in front of the vertebral column by a
fold of the peritoneum, and which consists of a collection
of small membranous sacs, rounded, more or less developed, and united in clusters. The walls of these sacs
abound in blood vessels, and secrete the ovules, which
are formed in their interior and which consist of a yellow material, enveloped in a very thin membrane. These
bodies slowly enlarge, and when they have acquired the
volume to be possessed by the yellow of the perfect egg,

the ovarian sac, in which each of them is contained, bursts
and allows it to escape into the cavity of the pavilion, a kind
of membranous tunnel applied upon the ovary and conducting outwardly by the oviduct, a tube of the same nature,
the inferior orifice of which is seen in the cloaca, near
the anus.   At the moment when the ovule descends into
the oviduct, it is merely composed of *vitellus*, or yellow,
enveloped in a membranous sac, at one point of which is
a whitish spot, called the *cicatricule*, and which deserves
notice, because in its interior the embryo is to be developed ; but, in proportion as the ovule descends, it is covered by other substances secreted by the walls of the
canal it traverses.   About the middle of the oviduct it is
enveloped by a thick and glairy matter, which is the
white of the egg, and a little lower down there is formed
around this new layer a thick membrane, the external
coat of which is finally encrusted by an earthy deposite,
and thus constitutes the shell of the egg.

In this condition the egg is laid.   When it has not
been previously fecundated, it does not undergo any important change ; but, in the opposite case, it becomes the
seat of an active progress as soon as its temperature is
suitably elevated.

If the cicatricule, which is about six millimetres in diameter, be now examined by the microscope, there will
be seen towards the centre a small white oblong body
which may be considered the rudiment of the germ, and
which presents a central white line rounded at its summit ; this spot marks the place in which the cerebro-
spinal cord will be developed, and, according to some
physiologists, it is itself the first appearance of the nervous system.   Around the germ, is seen a kind of mem-

branous and transparent disc which in its turn, is bordered by a darker zone and by two concentric light circles. About the eighteenth hour of incubation, the germ contracts, takes nearly the form of a lance, is rounded at its superior part and forms a fold, which doubles like a cloth in front of the cephalic extremity of the cerebro-spinal line. Upon the sides of this longitudinal line may also be remarked two small ridges, which enclose it as in a canal. Soon after these ridges unite by their inferior extremities and approach so as to conceal the line which separates them; finally, about the twenty-fourth hour, three pairs of rounded points make their appearance which are the first rudiments of the vertebræ, the number of them now rapidly increasing.

The transverse fold alluded to upon the anterior extremity of the germ, is the first rudiment of the head, which soon tends to detach itself and become distinct. Toward the thirty-sixth hour of incubation the eyes of the pullet.may be perceived; soon after, the posterior part of the body is plainly defined, and the embryo curves upon itself. During the third day the head becomes more and more distinct; its pointed extremity which corresponds to the beak, is bent upon the chest, and on the sides of the vertebral column under the form of small white tubercles may be recognised the earliest traces of the upper extremities; soon after, the inferior limbs are formed in the same manner; two small appendages fixed below the neck also appear, and constitute by their development the lower jaw; finally, the eyes are colored black. The fifth day of incubation, the limbs, which as yet resemble mere stumps without form, begin to execute some slight motions, and twenty-four

38

hours after they are so far developed that the thighs can
be distinguished from the legs, and the fore-arm from the
arm: the general form of the individual also approaches
a little to what it will be in future ; about this period the
heart enters the cavity of the chest, and the walls of the
abdomen are completed.   The seventh day the feet are
formed, and about the ninth day, upon the skin of the
embryo, may be found little pores, which are the openings of the capsules destined to secrete the feathers,
which begin to make their appearance at the end of the
tenth day, and cover the whole body in the space of
twenty-four hours.   The size of the head, at first exces-
sive, diminishes proportionally to that of the rest of the
body, and the eyes, which were remarkable for their size,
increase more slowly than the other parts ; the limbs, on
the contrary, are developed with greater rapidity, so that
the entire chick approaches more and more nearly the
perfect animal.

Considered with regard to its external form merely,
the embryo presents, as we have seen, true metamorphoses ; but the most curious part of the history of its
development is that which displays the manner in which
the different sets of apparatus, most important to life, are
successively formed in the interior of its body.

About the twenty-seventh hour of incubation, on the
anterior face of the pullet and precisely at the point where
the membrane folded in front of the head terminates,
may be perceived a small transverse cloud, which enlarges
at its two extremities, and is insensibly lost upon the
transparent area in the middle of which the germ is
placed.   This cloud is the rudiment of the left auricle of
the heart.   Three hours after the centre of this organ

will be found surmounted by a straight vessel directed towards the head, and which is the left ventricle ; soon after, a third enlargement is seen above the latter, it is the bulb of the aorta which afterwards disappears, but is permanent in certain reptiles such as frogs ; the heart then enlarges and curves upon itself ; a contraction is established between the auricle and the ventricle, and, about the thirty-sixth hour, the former of these cavities begins to ascend to the summit of the apparatus : at this period the heart begins to beat, but it does not yet contain blood, and is only filled by a colorless liquid. From the earliest hours of incubation, the transparent area around the germ also presents important modifications : the membrane forming it is divided into two leaves between which is developed a layer of spongy tissue which about the thirtieth hour begins to thicken in certain places, and to take a yellow tint ; this tissue gradually extends over the whole surface of the yellow, and little islets filled with a reddish liquid form in its interior ; finally these soon communicate together and form a vascular net-work, which surrounds the embryo and sends the blood to the heart by two vessels, the extremity of which is lost in the left auricle. The blood is at first formed in this vascular membrane and far from the embryo ; and when it begins to appear, its globules are circular. The circulation is then easily traced : the blood passes through the ventricle, arrives in the bulb of the aorta, thence enters the descending aorta, which soon divides into two branches, which issue from the body of the fœtus and are lost in the vascular area by which it is surrounded ; the blood which thus leaves the chick on the right and left, is divided into a collection of capillary

vessels, then runs into a general vessel which conducts it
upward or downward, whence it returns to the heart.
Between the third and fourth day of incubation, the right
ventricle may be clearly distinguished and appears as a
small sac placed in front of the left ventricle, communicating freely with the cavity of the auricle and continuous with a vessel, the extremity of which is directed
towards the point occupied by the lungs.   On the second
day, the right auricle also begins to be formed by means
of the development of an annular fold, which divides the
left auricle into two distinct parts.   Finally, about the
sixth day, we begin to perceive in the blood elliptical
globules, and the ninth day these have taken the place
of the circular globules which at first existed alone ;
their appearance coincides with that of the liver, and
with the obliteration of the vessels of the membrane of
the yellow, in which we found sanguification to commence ; therefore there is reason to consider this viscus
as the seat of secretion of these corpuscles.

The lungs begin to be developed about the fourth day ;
at first they consist of two oblong, almost transparent
tubercles placed behind the heart ; they soon acquire a
reddish tint, but do not serve for respiration till the chick
has broken its shell.

This function is however very actively executed from
the earliest moments of incubation ; and if air be prevented from penetrating the egg, the chick dies almost
immediately.   At the moment when laid, the egg is
completely filled by the white and the yellow, but these
liquids gradually lose by evaporation a certain quantity
of their water, and there is thus formed under the shell
a vacuum filled with air : at the same time the yellow

undergoes important modifications which render it lighter than the white, so that it occupies the superior part of the egg whatever be its position ; and the serosity, which accumulates during the second day of incubation under the cicatricule, producing the same effect upon this, causes it to float so as to be in contact with the air spoken of. The respiration of the embryo is at first effected by the contact with the air which has thus penetrated beneath the shell, or by the membrane of the yellow ; but soon after, this function becomes the duty of a new membrane called *allantois.* This begins to make its appearance about the forty-fifth hour of incubation, under the form of a membranous and transparent vesicle, of the size of a pin-head, placed in the abdominal region of the chick. This sac is rapidly developed, expands upon the superior surface of the yellow, and finally occupies the whole internal surface of the shell against which it is placed. Lastly, its external leaf is quickly covered by a magnificent vascular net-work, which receives the venous blood coming from the embryo, and places it in contact with the air to be transformed into arterial.

The intestinal canāl appears to originate by two folds of the internal layer of the cicatricule, which at first resemble tunnels open at one end, situated above the vertebral column and opposite each other; these folds gradually contract and close ; but their cavity still remains in communication with the yellow, which little by little enters it and serves to nourish the fœtus ; thus it is more and more diminished, and towards the end of incubation it is introduced into the interior of the abdomen.

Finally, the nervous system undergoes, in its development, a series of modifications yet more remarkable than

those we have observed ; and the transitory forms seen
in it have the greatest analogy with the permanent condition of these same parts in animals less elevated in the
zoölogical series.

The majority of animals have, on coming into the
world, nearly the forms and the mode of organization
they are to preserve during the whole of life, but it is
not so with all ; there are many which, even after birth,
pass through changes analogous to those already experienced during the development of their embryo, and
sometimes these changes are so complete that the animal
undergoes true metamorphoses before reaching the perfect state.   Frogs and especially insects furnish remarkable examples of these transformations.

APPENDIX.

PAGE 25, LINE 8.

The value of the millimetre according to the best authority being only .03937 portion of an inch, we can conceive, that it must require a microscope of no small power to display these corpuscles of the blood; those of the blood of man being only .00026 of an inch in diameter according to this calculation. For a description of the globules and of the peculiar properties of the vesicle, and of the nucleus, the reader may consult the recent works upon the Blood; those of Magendie, Schultz, Andral, Ancell, etc.

PAGE 26, LINE 20.

The non-coagulability of the blood as resulting from electricity has been denied by some writers. The following account is given by Scudamore in his Essay upon the Blood. "Experiment LII. A strong mongrel dog was the subject of the experiment. By means of a very powerful electrical battery, constructed by Mr. Woodward at the Surry Institution, the animal was killed after several shocks. On inspection next morning, the blood in the cavities of the heart, and in the vena cava, was found quite coagulated.

"I made a similar experiment with rabbits, with the same result; and in these smaller animals, death was effected instantaneously."

Page 29, Line 27.

Transfusion, as originally performed, is an instance of the gross errors into which practice, unsupported by theory, may fall.

"The operation of transfusion," says Dr. Palmer, in his notes to Hunter on the Blood, "was first performed on the human subject in France, by MM. Denis and Emmerts, on the fifteenth of June, 1667, and repeated in England by Drs. Lower and King on the twenty-third of November of the same year. It speedily grew into favor; but in consequence of the fatal and dangerous effects which ensued in several cases, it was interdicted by a sentence of the Chastelet, and soon afterwards fell into desuetude."

Page 57, Line 20.

In an essay upon the Blood by W. Stevens this doctrine of Endosmosis is fully considered, and some sources of error on the part of Dutrochet clearly pointed out. The passage is too lengthy to be extracted, and therefore my readers had better consult the work in question.

The value of the centimetre is .39371 of the inch.

Page 84, Line 28.

The value of the litre being 61.028 English cubic inches, it follows that the average consumption of oxygen by individuals would be not far from 45771 cubic inches, or 1585 pints per day. While the consumption of atmospheric air for the same space of time would amount to 213598 cubic inches, rather more than 7398 pints.

Page 95, Fourth Paragraph.

The apparent contradiction in this paragraph to the closing line of page 84 arises from the fact, that in the present instance the author considers the use of the atmospheric air, and in the other case the actual expenditure of this fluid was stated.

PAGE 98.

Many experiments upon the subject of animal heat with the view of determining its nature, cause, etc., have been made within a few years by the physiologists of Europe; among the most interesting are those by Dr. Alison, and by Edwards, (the brother of the author), detailed in his work upon the effects of physical agents upon life.

PAGE 106, LINE 25.

The following description of the plant bearing the cashew-nut is taken from the United States Dispensatory of Messrs. Wood and Bache.

"Anacardium Occidentale, Linn. *Cassuvium pomiferum*, Lam. *Cashew-nut.* A small and elegant tree, growing in the West Indies, and other parts of tropical America. A gum exudes spontaneously from the bark, which bears some resemblance in appearance to gum Arabic, but is only in part soluble in water, and consists of proper gum and bassorin. It is the *gomme d'acajou* of the French writers. The fruit is more important. It is a fleshy, pear-shaped receptacle, supporting at its summit a hard, shining, ash-colored, kidney-shaped nut, an inch or more in length, three quarters of an inch broad, consisting of two shells, with a black juice between them, and of a sweet oily kernel within the inner shell. The receptacle is of a red or yellow color, and of an agreeable sub-acid flavor with some astringency. It is edible, and affords a juice which has been recommended as a remedy in dropsy. This juice is converted by fermentation into a vinous liquor, from which a spirit is obtained by distillation, much used in making punch, and said to be powerfully diuretic. The nuts are well

39

known under the name of *cashew-nuts.* The black juice contained between their outer and inner shell, is extremely acrid and corrosive, producing, when applied to the skin, severe inflammation, followed by blisters or desquamation of the cuticle."

PAGE 107, LINE 9.

A beautiful instance in our own language is Byron's admirable picture of the Prisoner of Chillon. It will not perhaps be considered amiss, to introduce in this place the passage of Dante referred to as given in the version by Cary.

> " When I awoke,
> Before the dawn, amid their sleep I heard
> My sons (for they were with me) weep and ask
> For bread. Right cruel art thou, if no pang
> Thou feel at thinking what my heart foretold ;
> And if not now, why use thy tears to flow ?
> Now had they wakened ; and the hour drew near
> When they were wont to bring us food ; the mind
> Of each misgave him through his dream, and I
> Heard, at its outlet underneath lock'd up
> Th' horrible tower : whence, uttering not a word,
> I looked upon the Visage of my sons.
> I wept not : so all stone I felt within.
> They wept : and one, my little Anselm, cried,
> ' Thou lookest so ! Father, what ails thee ? ' Yet
> I shed no tear, nor answered all that day
> Nor the next night, until another sun
> Came out upon the world. When a faint beam
> Had to our doleful prison made its way,
> And in four countenances I descry'd
> The image of my own, on either hand
> Through agony I bit ; and they, who thought
> I did it through desire of feeding, rose
> O' the sudden, and cried ' Father, we should grieve
> ' Far less, if thou would'st eat of us ; thou gav'st
> ' These weeds of miserable flesh we wear,
> ' And do thou strip them off from us again.'
> Then, not to make them sadder, I kept down

My spirit in stillness. That day and the next
We all were silent. Ah, obdurate earth !
Why open'dst not upon us ? When we came
To the fourth day, then Gaddo at my feet   ·
Outstretched did fling him, crying, ' Hast no help
' For me, my father ! ' There he died ; and e'en
Plainly as thou seest me, saw I the three
Fall one by one 'twixt the fifth day and the sixth :
Whence I betook me, now grown blind, to grope
Over them all, and for three days aloud
Call'd on them who were dead. Then fasting got
The mastery of grief." — CANTO xxxiiii.

PAGE 110, LINE 1.

The term ' cul-de-sac ' may be applied to any cavity destitute of an outlet, although provided with an entrance.

PAGE 118, LINE 7.

This period is not entirely free from danger and calls for
the lancet to avert many secondary evils. Many a babe has
fallen a victim to hydrocephalus from the fear of the gum-
lancet on the part of the mother. In all diseases of infancy
there is so frequently some complication with the cutting of
the teeth, that it is by no means an unwise practice to put a
finger in the mouth for its examination.

PAGE 124, LINE 7.

The common expression, " the palate 's down," means
neither more nor less than that from a relaxed state the uvula
rests upon the base of the tongue, where by its irritation a
constant hacking, tickling cough is kept up.

PAGE 129, LINE 20.

Probably the most remarkable series of experiments ever
performed upon the living subject were those of Dr. Beaumont

upon Alexis St. Martin. The latter individual was wounded
by the bursting of a gun in such a manner that a perforation
into the stomach took place ; after a long period the wound
cicatrised and healed, leaving the opening closed merely by a
valve which could be raised and the whole action of the stomach brought into view — pieces of food could be introduced,
gastric juice extracted, etc.

Page 142, Line 3.

This important question is still mooted, but will not probably long remain so; the attention of physiologists is chiefly
directed to its adjustment at the present time.

Page 152, Instinct.

Cases now and then occur, in which this faculty seems
closely allied to an effort of the reasoning powers, but however near, there is no evidence that an animal is capable of
reflection — it may take cognisance of outward events, but it
lacks the capacity of investigating the mind as such.

Page 154, Line 3 — 16.

This comprehensive law of development is thus beautifully
expressed by Solly in his work upon the brain. " The mature
human frame, which in its perfect adaptation to fulfil the ends
of its existence strikes the philosophical anatomist with admiration, does not result from the gradual increase of an exact
though minute representation of its perfect form; but during
the course of its development, and while gradually progress-
ing towards its ultimate perfection, its constitution temporarily
assumes many forms, which are permanently retained by one
or other of the members among the lower orders of creation."
p. 226.

PAGE 165, LINE 24.

Because nerves from the ganglionic system are sent to the vessels, it has been inferred, that the blood itself was under the influence of this system.

PAGE 171, LINE 20.

M. Magendie only confirmed by his experiments what Sir Charles Bell had previously demonstrated by actual experiment to be the case.

PAGE 194, LINE 12.

An operation, from which some benefit has been derived in deafness, consists in the introduction into the outlet of the eustachian tube (*n*, fig. 31), of a minutely perforated tube, through which air is thrown by a machine into the cavity.

PAGE 211, LINE 13.

The contraction of the internal rectus muscle gives rise to strabismus or squinting ; for the relief of this deformity Professor Dieffenbach of Berlin proposed the division of the muscle. As the subject possesses so much interest 1 will extract from the Am. Jour. of Med. Science for August, 1840, an account of his mode of operating.

" The subject of this operation was a child seven years old, whose eye was drawn far into the inner angle of the eyelids so as to produce considerable disfigurement. The operation was performed in the following manner : — The head of the child was held against the chest of one assistant, while another with two hooks kept the eyelids widely apart. The operator then passed a third hook, which he gave to a third assistant to hold, through the conjunctiva, and to some depth in the subjacent cellular tissue at the internal canthus. He next fixed a fine double hook in the sclerotica at the inner angle, and, taking it in his left hand, drew the eye outwards. Then cutting into the conjunctiva close to the ball, where it is continued from it to the internal canthus, and penetrating more deeply by separating the cellular tissue by the side of the sclerotica, he divided the internal rectus muscle close to its insertion with a fine pair of scissors. The eye was immediately drawn outwards by the external rectus, as if it had received an electric shock; and in another instant became straight, so that there was no difference perceptible between its direction and that of the other eye."

He has been followed by several, who have variously modified his mode of operating, not only in Europe but also in this country. In the Boston Med. and Surg. Journal, December 2d, 1840, are six cases of the operation as performed by J. H. Dix, M. D., among the earliest cases to be found in the surgical annals of our country.

Page 212, Line 31.

If each eye has its own optic nerve, why do we not perceive two images?

"The rays of light from any object, placed laterally, impinging upon the retina of both eyes, will strike the outer side of one eye and the inner side of the other; now, * * * * it follows as a necessary consequence, that the outer and inner side of each opposite retina is formed by one and the same nerve, a peculiarity of structure that goes far to account for the circumstance so often reasoned upon, namely, that a single impression is conveyed to the sensorium, though each eye receives the impression. Whether this mode of accounting for it be satisfactory or not, the following facts are extremely interesting, and not sufficiently known, namely, that in those fishes whose eyes are placed so completely on the side of the head that the rays of light from any given object cannot impinge on both retinæ, as, for instance, in the cod and haddock, the optic nerves, instead of forming any union or commissure, cross each other completely, having a membrane interposed between them : in those fishes, again, whose eyes are situated so that even a small portion of their retinæ correspond, as in the carp, we find a few commissural fibres; and in those whose retinæ correspond in every point, as in the skate, we find the commissure as complete as in the human being." — *Solly on the Brain*, p. 243.

PAGE 220, LINE 2.

No less than seven bones enter into the formation of the orbit; superiorly, the frontal and sphenoid; externally, the same, with the malar; inferiorly, the palate, superior maxillary, and malar; internally, the frontal, superior maxillary, sphenoid, ethmoid and lachrymal.

PAGE 232.

It is rather too late to discuss the question whether phrenology be true or false, the researches of M. Flourens in many cases support the doctrine. Much confusion has been no doubt caused by considering the brain as the cause rather than the instrument; by regarding the brain as the mind, and not as the agent through which the mind acts.

PAGE 237, LINE 25.

The theory that muscular contraction is due to electric currents has been long held in high estimation by the German physiologists; the following rough translation from a work by Joh. Christian Heinroth upon Anthropologie, Leipzig, 1822, embraces the most important points.

"As the charging of the nerves was considered an electric act, so also the discharging of the muscles; the nerves were to convey sensation, but the muscles motion. The manifestation of motion by the muscles is one of the deepest mysteries of nature, as mysterious as that of sensation by the nerves. With the speed of lightning is the muscle put in action, a proof of the electric nature of the motion; it is no mechanical, but a higher power, active in the muscles. The latter merely convey the expansive, or motor power; they are, as it were, permeated by this power; an electric stream courses through them. The most wonderful is their harmonious motion; the muscles are moved in groups — a proof that the impulse to motion arises from a single point and is dispersed like rays to the various muscles. This One-dom point cannot exist in the muscles themselves, it must be in the exciting organs, the nerves. The charged nerve sends the electric stream to the muscles, which discharge themselves in groups in swift succession and with many regular, sometimes rectilinear, at others curvilinear motions."

Page 244.

The following curious experiment by Dr. Hunter, illustrates most clearly the manner of growth in the bone.

"I took a pig of a very large breed when young, bored two holes in the tibia, and put a shot into each, measuring on a card the distance of each from the other. I allowed this pig to grow up to its full size, then killed it and took out the bone, and I found the two holes at exactly the same distance from one another as at first. Now if the bone had grown in all its parts these two shot would have been removed to a distance from each other proportionate to the growth of the bone."

Therefore, " bones do not grow by having new particles put into the interstices of previously formed parts, so as to remove these to a greater distance from each other, by which means they should grow larger, — as, for instance, if I put a sponge into water, the water getting into all the interstices makes it larger, — but they grow by the addition of new bone on the external surface." — *Hunter's Prin. of Surg.* p. 48.

Page 280, Line 29.

In most cases of attempted suicide the opening is made so high up as to defeat the object, and almost without affecting the voice.

VALUABLE WORKS

RECENTLY

PUBLISHED, IN PRESS, AND FOR SALE

BY

CHARLES C. LITTLE AND JAMES BROWN,

BOOKSELLERS, IMPORTERS, AND PUBLISHERS,

NO. 112 *WASHINGTON STREET....BOSTON.*

JANUARY, 1841.

Books lately Published.

HISTORY OF THE UNITED STATES FROM THE DISCOVERY OF THE AMERICAN CONTINENT. In 3 volumes. By GEORGE BANCROFT. The FIRST PART of this work, embracing the HISTORY OF THE COLONIZATION OF THE UNITED STATES, is now completed. It forms three volumes 8vo, and contains the account not only of the settlement of the thirteen original states, but of the Spanish settlements in Florida. and of the French discovery and colonization of Michigan and Wisconsin; the discovery of the Mississippi, the colonization of Illinois and Indiana, of Mississippi and Louisiana, and the attempts at colonizing Texas by La Salle. The topics most interesting to the people of the great valley of the Mississippi are delineated more fully than in any American work, and from original sources.

The book is printed in the best style, equal to that of the London press, and is richly illustrated by maps, sketches, and engravings, particularly by heads of the Winthrops, of Smith, of William Penn, and Franklin, fac-similes of the first maps of the valley of the Mississippi, and of Lake Superior, with sketches illustrating Indian life and appearance.

This work has been favorably noticed in some of the best journals of Germany and England, and in the chief American periodicals.

"We know few modern historic works, in which the author has reached so high an elevation at once as an historical inquirer and an historical writer. The great conscientiousness with which he refers to his authorities and his careful criticism, give the most decisive proofs of his comprehensive studies He has founded his narrative on contemporary documents, yet without neglecting works of later times and of other countries His narrative is every where worthy of the subject. The reader is always instructed, often more deeply interested than by novels or romances. The love of country is the Muse which inspires the author; but this inspiration is that of the severe historian, which springs from the heart."—*Göttingen Review, written by the celebrated historical Professor* HEEREN.

"If any truth ever was 'stranger than fiction,' it will be found in this history; so varied, so vived, so heterogeneous, and yet so harmonious a picture do these early American annals set before us; so poetical are they, though at the same time so notoriously and preëminently practical —in a word, as we began with saying, so picturesque.

" Numberless illustrations of the truth of this general remark might easily be der
from Mr. Bancroft's pages. He is poet enough to have clearly discerned and fully
preciated the splendid, poetical·interest of his theme, and to have given to its var
passages accordingly the prominence they deserved. He does not, however, forge
neglect his business in his delight ; he only avails himself of his excitement to do
business better ; he uses his enthusiasm as a minister to his industry. We rejoice to
meanwhile, that the work is appreciated in America as we hope it will be in England
London Athenæui

" Mr. Bancroft, who is an American himself, possesses the best qualities of an histo
His diligent research, his earnest yet tolerant spirit, and the sustained accuracy
dignity of his style, have been nobly brought to bear upon one of the grandest sub
that ever engaged the study of the philosopher, the legislator, or the historian. T
can be no doubt of his being possessed of the highest requisites of an historian."
London Monthly Revie

" On the score of research, correct conception of the beautiful in nature and in hu
character, and profound learning, united to copiousness and splendor of diction, this
uable production is entitled to the highest meed of praise."—*Canada Times, Dec.* 1

" A History of the United States, by an American writer, possesses a claim upon
attention of the strongest character.
" It would do so under any circumstances, but when we add that the work of
Bancroft is one of the ablest of the class, which has for years appeared in the En
language ; that it compares advantageously with the standard British historians ; tha
far as it goes, it does such justice to its noble subject, as to supersede the necessity of
future work of the same kind ; and if completed as commenced, will unquestionably
ever be regarded, both as an American and as an English classic."—*Review in the N
American, by Gov.* EVERETT.

" The favorable notice we took of Mr. Bancroft's labors on his first appearance has
fully ratified by his countrymen. His colonial history establishes his title to a ɪ
among the great historical writers of the age."—*Review in the North American, Jan.* 1

" We speak with confidence when we say Mr. Bancroft's persevering scrutiny of
torical memoirs and documents, and the skilful use he has made of them, are calcul
to give to his works, permanent and universal value, as a classical history of the Uɪ
States."—*Philadelphia American Quarterly Review.*

" Gladly would we extend these extracts, but we have already exceeded our s
Most heartily do we commend the history to our readers. We commend it, espec
to YOUNG MEN. No more profitable volumes can they study. It should not be th
proach of any of their number that they are ignorant of the past annals of their couɪ
With such attractive volumes as those of Mr. Bancroft at command, no excuse ca
valid for such neglect."—*Boston Morning Post,* 1840.

" A good, and at the same time a copious history of the United States. Mr. Ban
has shown himself admirably qualified for the undertaking, by an easy and flowing s
patience of research, and faithfulness of delineation."—*New York Journal of Comm*

" We consider it a source of congratulation to the whole nation, that so accompli
a scholar, so patient an investigator, and so eloquent a writer, has undertaken the n
needed task of writing a worthy history of these United States. In the volume be
us, we see abundant evidence that, while truth will—at any expense of labor in ferre
it out from the original authorities, instead of relying, as is so common, upon the c
of copies—be fearlessly spoken, no prescription of time or great names will be alle
to sanction error.   *   *   *   It will be received, we feel well assured, as a wo
offering to his country, from one of her able and qualified sons."—*New York Amer*

" These volumes are eagerly sought and read. The characteristics of Mr. Bancɪ
history are, patient research, apt illustration, original and felicitous grouping of chara
and events, and general impartiality. His style is ornate and elaborate ; always arreɪ
and retaining attention."—*New York American, Dec.* 1840.

" It is a matter of question to us, if any history in the language contains more b
tiful touches. A mine of wealth is here opened to the artists of our land ; and w
much at fault, if the beautiful sketches and touching outlines of the author, are not
transferred to the canvass. Most sincerely do we trust, that these volumes ma
placed in all our houses, and that our young men may become intimately versed in
pages of this remarkable production."—*New York Commercial Advertiser, Jan.* 1841.

THE CATHOLIC. By William H. Prescott.

WITH SPLENEID PORTRAITS AND MAPS.

In 3 volumes 8vo. 7th edition..

In this edition the publishers have added a Map for the War of Grenada and Gonzalvo de Cordova's Campaigns in Italy.

" The history of Spain cannot boast of a more useful and admirable contribution si the publication of the great work of Robertson."—*British and Foreign Review.*

" Bold indeed it is; but in our judgment eminently successful. On such works are content to rest the literary reputation of the country."—*North American Review.*

" In every page, we have been reminded of that untiring patience and careful disci ination, which have given celebrity to the great, though not always impartial, histo of the Decline and Fall of the Roman Empire."—*New York Review.*

" His laborious industry, conjoined with native ability, places the author next HALLAM, amongst our living historians, for the largeness and philosophical justnes his estimate, the distinctness and comprehension of his general surveys, and the in esting fulness of his narrative."—*London Spectator.*

" Mr. Prescott's is by much the first historical work which British America has produced, and one that need hardly fear a comparison with any that has issued from European press since this century began.—*London Quarterly Review.*

"One of the most remarkable historical compositions that have appeared for a l time."—*Bibliotheque Universelle de Geneve.*

"Mr. Prescott's merit chiefly consists in the skilful arrangement of his materials the spirit of philosophy which animates the work, and in a clear and elegant style t charms and interests the reader. His book is one of the most successful historical ductions of our time. The inhabitant of another world, he seems to have shaken off the prejudices of ours. In a word, he has, in every respect made a most valuable a tion to our historical literature."—*Edinburgh Review.*

THE WORKS OF EDMUND BURKE, IN NINE VOLUMES, OCTAVO

" The present edition of Burke's works is more complete than any one which hitherto appeared either in England or America It comprises the entire contents of English edition of his works in sixteen octavo volumes, including two volumes of speec on the trial of Hastings, published in 1827, and which have never before been republis in this country. It also contains a reprint of the work entitled ' An Account of European Settlements in America, first published in 1761, which, though publis anonymously, is well known to have been written by Burke, and also the corresponde with French Laurence, which are not contained in the English edition of his collec works. Although the present edition contains a volume more than the latest and English one, it is offered at less than one half the price. It has been the aim of the p lishers to present the work in a form and style worthy of its contents; and it is confider offered to the favorable regard of the public from its completeness, its moderate pr and its typographical excellence."—*Preface to the Edition.*

" The publishers of this edition have borne in mind the nature and value of the c tents, and given the mechanical execution a degree of excellence corresponding to literary and political merits of the volumes, of which there are nine, in octavo, cont ing all that is given in the sixteen octavo volumes of the London edition, and includ two volumes on the East India question, and an account of the European discovery America, which last is not in the London edition.

" This edition of Burke may be considered as containing all that that distinguis writer intended for the press, and is commended to the taste and judgment of those are forming or completing a library.—*United States* (Philadelphia) *Gazette.*

. " Possessed of most extensive knowledge, and of the most various description; quainted alike with what different classes of men knew, each in his own province, with much that hardly any one ever thought of learning; he could either bring masses of information to bear directly upon the subjects to which they severally belo ed—or he could avail himself of them generally to strengthen his faculties and enla

his views—or he could turn any portion of them to account for the purpose of illustra
his theme, or enriching his diction. Hence, when he is handling any one matter,
perceive that we are conversing with a reasoner or teacher, to whom almost every o
branch of knowledge is familiar. His views range over all cognate subjects; his rea
ings are derived from principles applicable to other matters as well as the one in ha
arguments pour in from all sides, as well as those which start up under our feet,
natural growth of the path he is leading us over; while to throw light round our st
and either explore its darker places, or serve for our recreation, illustrations are fetc
from a thousand quarters; and an imagination marvellously quick to descry unthou
of resemblances, pours forth the stores, which a lore much more marvellous has gath
from all ages, and nations, and arts, and tongues. We are, in respect to the argum
reminded of Bacon's multifarious knowledge, and the exuberance of his learned far
while the many-lettered diction recalls to mind the first of English poets, and his
mortal verse, rich with the spoils of all sciences and all times."

Lord Brougham.—Sketches of English Statesme

THE POETICAL WORKS OF EDMUND SPENSER, in 5 volumes, 12
first American edition, with introductory observations on the Faery Queene,
notes by the Editor.

A few copies beautifully printed on large paper, octavo.

☞ Copies in fine binding for presents, &c.

The character of this edition will be learnt by the letters and notices which foll

From Professor Ticknor—Boston.

Messrs. LITTLE & BROWN,

Gentlemen. Sir Walter Scott, in a Review of Todd's edition of Spenser, written
dently with kind feelings towards its editor, cannot help saying, at the end of it, "
conclude with a single hint. Mr. Todd is a man of learning and research. We wis
would write essays in the Archæologia and renounce editing our ancient poets." Tc
edition, however, is the only one that can come into competition with yours; since, v
was done in the middle of the last century by Upton, Church, and Warton for the F
Queene, and by Hughes for the whole works of Spenser, is all used by Todd. The
no doubt, therefore, you have published the best edition of Spenser yet known. But
have, I think, done more than this. You have, it seems to me, published a positi
good, useful, and agreeable edition of him; one that will cause him to be read and
joyed by many classes of persons, who would otherwise not have ventured to oper
pages. I have been in the habit of using Todd's edition these twenty years, and
winter read in it all Spenser's poetry; and as I have recently gone over nearly the w
of your edition, I feel as if I could judge fairly of its value. The result is, a strong
suasion in my mind, that the talent and taste shown in the beautiful introductory ol
vations of your accomplished editor, with the short, exact, and sufficient notes, gloss
and explanatory, which he has put at the bottom of each page, where they are wai
and not at the end of the work, where they would be almost useless, constitute this
lication a real and permanent service rendered to the cause of English literature in
country; a service the more important, as, while you have made your edition good,
have also made it typographically attractive, and yet so cheap that few readers of En
literature need to refuse themselves the pleasure of owning it. I wish you would
lish similar editions of Chaucer, and of the old Ballads, and the other old poetry scati
in the collections of Percy, Ritson, Evans, Scott, Hartshorne, &c. Let me add, tl
am happy to believe there is encouragement for such undertakings among us; and
the public begin to buy good and tasteful editions of our great English poets, in pla
the pretty "Annuals," as they are called, which are, in general, only beautifully c
mented trash.

Your obedient servant, GEORGE TICKNC

From the Author of Ferdinand and Isabella.

Messrs. LITTLE & BROWN,

Gentlemen. I have examined the copy of Spenser which you have put into my ha
and am very ready to bear my testimony, for as much as it is worth, to the excellent r
ner in which the edition has been prepared. A good edition of this old bard is almo
difficult as that of an ancient classic. The poets of his age abound in such local
temporary allusions, as are most natural in a period, when the taste is less cultiv
than the imagination. In addition to these, Spenser enveloped himself in metaphor
allegory, of the most complicated kind, while he affected a language, regarded as
quated even by his contemporaries. Owing to these circumstances, he is, although
most picturesque writer in the most picturesque age of English literature, less read

many other poets, of inferior merit, some of whom, it may be added, have known how to cover up their own nakedness with his fine things, without much risk of detection. It was fortunate, that the business of editing this poet in the popular form which you intended, should have been committed to a man, who combines, what is rare,—scholarship, with a quick perception of the beautiful, and a good taste which has made him more anxious to show off his author than himself. There is no pedantic parade of learning in the notes, which are directed to instruct the reader in what is obscure in the text, or to point out its less obvious peculiarities Every one will appreciate the value of this, who has waded through the learned lumber of Spenser's preceding commentators, under whose hands the flowers of poetry seem to wither, while they are digging and delving after the roots from which they have sprung. In addition to these excellent Notes the text is further illumined by a glossary of the obsolete words, judiciously placed at the bottom of the page, where it may be embraced at a single glance with the text, an arrangement far preferable to that of other editions, where the glossary, from being removed to the end of the Canto, or the volume, imposes a task on the reader, little less burdensome than that of working his way with a dictionary. The "Introductory observations to the Faery Queene" are a very valuable addition to the stores of literary criticism, exhibiting a nice discrimination of the characteristics of the poem, with a genuine sensibility to its extraordinary beauties, which will not surprise those who are familiar with the writings of the accomplished Editor. It is needless to remark on the excellence of the typography and the mechanical execution of the work, as they are obvious to all. I heartily wish it may find the patronage, which so liberal an undertaking merits.

Very truly, your obedient servant,   WM. H. PRESCOTT.

"Messrs. Little & Brown, of Boston, have published, in five volumes, the entire works of Spenser, with a Biography of that eminent poet. They have done full justice to the author in the handsome style in which they have caused the work to be issued We are glad to see this spirit of liberality in the publishers, and doubt not that it will be met by a correspondent willingness on the part of readers to add slightly to the expense, in order to insure good appearance."—*United States Gazette.*

"The appearance of a beautiful edition of Spenser, with an original introduction and notes by an eminent editor is an event of no trifling importance in the annals of literature. In the second book of the Faery Queene the poet cites the wonders of the new discovered Western World, as a proof that in describing the marvels of Fairy land he is not exceeding the bounds of possibility. Could he have foreseen that in little more than two centuries the verses he was penning, would be published, commented and lectured upon, and familiar as household words in that new found Atlantis, which seemed to him fit to be compared with the fabulous Elfin kingdom, he would perhaps have been ready to confess that all the splendors which his boundless imagination had conjured up were to be eclipsed by the amazing realities which time was to bring into being.

"A grateful posterity has set up statues and storied monuments to the memory of the mighty dead; but we regard this tribute of respect and admiration for Spenser, as a nobler monument of his worth than the marble which Essex raised over his grave in Westminster Abbey. The simple, but dignified inscription on that marble informs the reader that Spenser's ' divine spirit needs no other witness, than the works he left behind him.' A monument has now been reared in a remote age and in a distant world which sacredly verifies these words.

"An edition like this has long been a desideratum in our libraries ; for we have hitherto had no other alternative in reading Spenser, but to use an edition with no notes at all, and no explanation but a glossary appended to the last volume, or an edition so encumbered and loaded down with notes and comments, that a fortune would be required to purchase it, and a life-time to read it. This want is supplied by the beautiful edition now offered to the public. The notes are few, discreet and interesting, but sufficient to explain the text ; and the glossary is printed at the bottom of each page just above the notes. The difference between reading with these aids, or reading an edition in which the glossary is printed at the end of the last volume, is nearly as great as we find between studying a Latin or Greek author, with an intelligible translation printed on the same page with the text, or searching for his meaning by the help of a lexicon alone.

"The Introductory Essay is an admirable critique, giving a comprehensive and satisfactory view of the outline of the poem, its peculiar beauties and merits, the historical references, and the nature of the allegory. It is written in a most engaging style, and displays a thorough acquaintance with the poem, and with all that has been written about the poem. Without any parade of learning the writer shows, in every line, that strength which learning alone can give."—*Boston Daily Advertiser.*

"We are glad to see that the old orthography is strictly preserved throughout, though we have heard some good natured people remark, that they should like to have had the spelling a little modernized! not reflecting that more than half the flavor of the different works would have evaporated in the process. They would still have contained much

good poetry, but it could hardly have been called the poetry of Spenser, for suc
change would have involved many others, still more important. We repeat, that
Hillard has discharged the duties which he undertook in a most excellent manner—cr
able at once to himself and useful to all his readers. We must not omit to mention,
these volumes, in point of mechanical execution, are among the most elegant that 1
ever come from an American press. They will bear a rigid comparison with the
elaborately finished London productions "—*Boston Morning Post.*

" We consider this reprint of Spenser, with the excellent annotations and other labo
the American editor, as an important step towards the more general introduction of w
of this class. It is remarked, properly, that the scholar will find only familiar matte
these volumes; but even such a reader will return to the Faerie Queene with rene
pleasure in the attractive dress of tasteful typography."—*Philadelphia National Gazet*

A MANUAL OF POLITICAL ETHICS, DESIGNED CHIEFLY FOR T
USE OF COLLEGES AND STUDENTS AT LAW ; PART FIRST, C
TAINING BOOK I. ETHICS GENERAL AND POLITICAL ; BOOK
THE STATE. By FRANCIS LIEBER. 2 vols. 8vo.

I beg leave to say, without meaning any formal compliment whatever, that y
Manual of Political Ethics is a profound work, full of deep reflection, solid principle,
sound and apposite illustrations. I have read it over superficially, but I have begun
have far advanced in the *study* of it with notes. I think your ethical and political p
ciples just and admirable and most instructive as to right, duties, property, social i
tions, government, sympathy, &c. &c., and I hope and intend to make myself fam
with your work as a text book Yours most truly,   JAMES KEN

" The work abounds with profound views of government, which are illustrated v
various learning. To me many of the thoughts are new, and as striking as they are r
If I may be allowed to judge of the whole work from this specimen, I do not hesitat
say that it constitutes one of the best theoretical treatises on the true nature and obj
of government, which has been produced in modern times, containing much for inst
tion, much for admonition, and much for deep meditation, addressing itself to the v
and virtuous of all countries. The work has with me a still greater value, in tha
aim and end are practical." &c.—*Extract from a letter of the Hon.* JOSEPH STORY.

 * * * " They are both (the State and Absolutism) subjects of deep importance,
are treated in a new and very interesting manner. With some, not important excepti
I am disposed to concur in the views which you take.

" I hope you will present the work to the public at an early day. It would be q
an acquisition to our literature, and do much to call public attention to subjects in w
all have so great an interest."—*Extract from a letter of the Hon.* JOHN C. CALHOUN.

" The various mistakes occasioned by the habit of seeking for the origin of societ
the rudest conceivable condition of human existence, falsely denominated the ' *sta
nature,*' are treated in a masterly manner in the Second Book of the work before us."

" Many other topics, discussed with ability and clearness and enriched with a g
variety of illustration, we cannot even refer to, without exceeding our limits. We t
that this work will be the means of calling the attention of our countrymen to a sub
in which they of all others, ought to be well versed."—*London Examiner.*

" Considering the systematic completeness with which the subject is handled, and
manner in which the accidents of things are put aside and the essentials aimed at, if
always reached, *Political Ethics* may be classed amongst the best treatises on Govern
since the days of *Aristotle.*"—*London Spectator.*

LEGAL AND POLITICAL HERMENEUTICS, OR PRINCIPLES OF
TERPRETATION AND CONSTRUCTION IN LAW AND POLITI
WITH REMARKS ON PRECEDENTS AND AUTHORITIES. Enlar
edition. 12mo. By FRANCIS LIEBER.

 * * * " I regard the Hermeneutics as a work eminently useful to our professic
not merely useful to students, but to men of long experience at the bar—as a most l
exposition of the principles and admirable illustration of the science of interpretation
construction. It is a valuable contribution to the law. I have given the Ethics on
cursory examination in MS. ; but from this rapid glance I am induced to believe, th
is well worthy the reputation of the author; and I cannot but think that the public

DICTIONARY OF LATIN SYNONYMES FOR THE USE OF SCHOOLS AND PRIVATE STUDENTS, WITH A COMPLETE INDEX. By LEWIS RAMSHORN; from the German, by FRANCIS LIEBER. 12mo.

" We are glad to see, in our own language, a translation of this valuable work of an eminent German scholar and practical instructer. If the Latin language is still to be a part of our course of education,—and we hope it will long continue to be so,—it must be studied with the aid of such works as the present; for which, indeed, we shall be obliged, for some time, to look to Germany, now at the head of the literature of all Europe.

" We cannot but congratulate the students of the Latin language in this country upon the publication of a work, which is superior to any one of the kind, that we are acquainted with, in the English language; and it cannot fail to be considered a necessary part of the *apparatus* of every student's library, as well as of every school where the Latin language is taught.

" We must not conclude our remarks upon this volume, without adverting to the extraordinary care with which it has been carried through the press; a consummation, not so easy as most readers would imagine, in works where the variety of types and languages is apt to mislead the most lynx-eyed corrector, and in school-books, above all others, of the highest importance."—*North American Review.*

MECANIQUE CELESTE, by the MARQUIS DE LA PLACE, translated by NATHANIEL BOWDITCH, L. L. D., to which is prefixed a Life of DR. BOWDITCH, with Portraits, &c. &c. Complete in 4 volumes, royal 4to.

A few sets of this great work, now recently completed, may be had if applied for soon. As it never will be reprinted, Public Libraries and individuals, who may wish to possess it, will do well to send their orders without delay.

MEMOIR OF NATHANIEL BOWDITCH, by his Son, NATHANIEL INGERSOLL BOWDITCH, originally prefixed to the 4th volume of the Mécanique Céleste. 1 volume 4to. 172 pages. With plates; splendidly printed.

A DISCOURSE ON THE LIFE AND CHARACTER OF THE HON. NATHANIEL BOWDITCH, L. L. D., F. R. S., delivered in the Church on Church Green, Boston. By ALEXANDER YOUNG.

THE FRENCH REVOLUTION. A HISTORY. Vol. I.—THE BASTILLE. Vol. II.—THE CONSTITUTION. Vol. III.—THE GUILLOTINE. By THOMAS CARLYLE. A new edition, corrected and enlarged, in 3 vols. 12mo. London and Boston. 1839.

" Of all books this is most graphic. It is a series of masterly outlines *à la Retzch.* Oh more, much more. It is a whole *Sistine Chapel of fresco à la Angelo*, drawn with a bold hand in broad lights and deep shadows. Yet again it is gallery upon gallery of portraits, touched with the free grace of Vandyke, glowing with Titian's living dyes, and shining and gloomed in Rembrandt's golden haze. And once more, let us say in our attempt to describe this unique production, it is *a seer's second sight of the past.* We speak of prophetic vision. This is a *historic vision*, where events rise not as thin abstractions, but as visible embodiments; and the ghosts of a buried generation pass before us, summoned to react in silent pantomime their noisy life."—*Boston Quarterly Review*, Oct. 1838.

A SYNOPSIS OF THE BIRDS OF NORTH AMERICA. By JOHN JAMES AUDUBON, F. R. S. S. L. & E. 8vo.

ORNITHOLOGICAL BIOGRAPHY, OR AN ACCOUNT OF THE HABITS
OF THE BIRDS OF THE UNITED STATES OF AMERICA; ACCOM-
PANIED BY DESCRIPTIONS OF THE OBJECTS REPRESENTED IN
THE WORK ENTITLED THE BIRDS OF AMERICA, AND INTER-
SPERSED WITH DELINEATIONS OF AMERICAN SCENERY AND
MANNERS. By JOHN JAMES AUDUBON, F. R. S. S. L. & E. 5 vols. Roy-
al 8vo.

A GRAMMAR OF THE ITALIAN LANGUAGE. By PIETRO BACHI, Instructer in Harvard University. A new edition revised and improved, with the addition of Practical Exercises and numerous Illustrations, drawn from the Italian
Classics. "Una lingua deve avere l' *uso* per base l' *esempio* per consiglio, e la
ragione per guida." *Cesarotti.* 1 vol. 12mo.

MEMOIR OF THE LIFE OF JOSIAH QUINCY, JUN. OF MASSACHU-
SETTS. By his son, JOSIAH QUINCY.

THE AMERICAN CONVEYANCER.
By GEORGE T. CURTIS, of the Boston Bar. 12mo.

This work contains all the varieties of legal forms of conveyancing and certify-
ing, that may be useful to the practical or the professional man, in all parts of the
United States. It contains a complete system of Forms and Directions for appli-
cants for Letters Patent; for the organization of Corporations; for the formation
of joint stock companies: and the copies intended for the use of this common-
wealth embrace the entire system of Insolvency under the Act of April, 1838. The
publishers believe that to both the lawyer and the layman it is a work of great
utility, avoiding many of the defects and improving upon many of the excellencies
of the books of conveyancing heretofore in use in this country.

LETTERS ON BOSTON. By E. C. WINES.

" This little volume is the best guide the stranger can put in his pocket when he visits
Boston. It points out every object worthy of notice in Boston and its vicinity; its schools
hotels. places of public amusement, architecture, libraries, with Letters on Mount Auburn,
Cambridge, &c ; interspersed with lively and judicious criticism by a gentleman whose
larger works have placed him amongst the most popular writers of the country."
Literary Telegraph.

THE INVENTOR'S GUIDE. COMPRISING THE RULES, FORMS, AND
PROCEEDINGS, FOR SECURING PATENT RIGHTS. By WILLARD
PHILLIPS. 12mo.

This Treatise embraces the laws and decisions, and principles and forms, that
were considered to be of practical importance to Inventors and Patentees; omit-
ting the legal proceedings and such other matters as were thought to be peculiarly
useful to members of the profession of the law.

DRAMAS, DISCOURSES, AND OTHER PIECES,
By JAMES A. HILLHOUSE. 2 vols. 16mo.

" About fifteen or twenty years ago, Mr. Hillhouse was well known as the author of
Percy's Masque, Hadad and other Poems, which gave proof of as much poetical talent,
as the country had exhibited, and placed him in the small number of its favorite and
most favored authors. But Mr. Hillhouse printed his own works in his own way; and
though they were speedily taken from the market by eager admirers, he neglected or re-
fused to permit further editions of them to be published, until, for many years past, they

have become absolute rarities. At last he has suffered them to appear again, adding to Hadad and Percy's Masque, Demetria, a tragedy of great beauty and power; and making a second volume of three prose Discourses, the Poem of Judgment, and the Sachem's Wood. It is long since so graceful and so truly poetical an addition has been made to the body of our literature. The style of their execution, too, is suited to their character. We count upon them, therefore, to do us credit in all respects, at home and abroad, for there is a finish about the poetry and graceful elegance thrown round its power, that are rare in any country."—*Boston Daily Advertiser.*

" Mr. Hillhouse is known to be one of the best writers of our country. There is a smoothness and finish about his productions, which we do not often see surpassed. Some of the Discourses are upon subjects of interest, which are treated with force and perspicuity."—*Boston Morning Post.*

SALLUST'S HISTORIES OF THE CONSPIRACY OF CATILINE AND THE JUGURTHINE WAR. FROM THE TEXT OF GERLACH, WITH ENGLISH NOTES. Edited by Henry R. Cleveland, A. M., late of the Boston Latin School. Stereotype edition. 12mo.

" The best and cheapest Edition of Sallust for Schools."

ORATIONS AND SPEECHES, ON VARIOUS OCCASIONS. By Edward Everett. 1 vol. Octavo.

A CONCISE APPLICATION OF THE PRINCIPLES OF STRUCTURAL BOTANY TO HORTICULTURE, CHIEFLY EXTRACTED FROM THE WORKS OF LINDLEY, KNIGHT, HERBERT, AND OTHERS, WITH ADDITIONS AND ADAPTATIONS TO THIS CLIMATE. By J. E. Teschemacher. Boston, 16mo.

" Every person engaged in Horticulture or Agriculture will find this little work a rich treasure."

THE TRIAL OF JESUS BEFORE CAIPHAS AND PILATE. BEING A REFUTATION OF MR. SALVADOR'S CHAPTER, ENTITLED "THE TRIAL AND CONDEMNATION OF JESUS." By M. Dupin, Advocate and Doctor of Law.

" If thou let this man go thou art not Cæsar's friend."—John xix. 12.

Translated from the French, by a Member of the American Bar. Boston, 16mo.

THE HISTORY OF NEW ENGLAND FROM 1630 TO 1649. By John Winthrop, Esq., first Governor of the Colony of the Massachusetts Bay. From his original manuscripts. With Notes to Illustrate the Civil and Ecclesiastical Concerns, the Geography, Settlement, and Institutions of the Country, and the Lives and Manners of the principal Planters. By James Savage. 2 vols. 8vo.

A NEW AND COPIOUS LEXICON OF THE LATIN LANGUAGE; COMPILED CHIEFLY FROM THE MAGNUM TOTIUS LATINITATIS LEXICON OF FACCIOLATI AND FORCELLINI, AND THE GERMAN WORKS OF SCHELLER AND LUENEMANN. Edited by F. P. Leverett. Royal 8vo.

PASSAGES IN FOREIGN TRAVEL. By Isaac Appleton Jewett. 2 vols. 12mo.

COLLECTIONS OF THE MASSACHUSETTS HISTORICAL SOCIETY. Complete. 27 volumes, 8vo.

PRINCIPLES OF THE THEORY AND PRACTICE OF MEDICINE.

MARSHALL HALL, M. D. First American edition, revised and much enlarged
JACOB BIGELOW, M. D., and O. W. HOLMES, M. D. 724 pages, 8vo.

This English work, by an author of great celebrity, has been revised and a
mented with new matter adapting it to the present state of Medical Science
the American editors. It appears from the advertisement that one third of
entire volume is written by the editors. The following are some of the opin
of the American press in regard to this Edition.

" We would unhesitatingly pronounce it the best and most complete text book for
study and practice of medicine. It is full of facts, well arranged and digested, and
from the endless repetitions and diffuse, ill digested matter which are often introd
into treatises upon medicine. The present state of the science is reached in almost e
instance."—*Philadelphia Medical Examiner.*

" In our next number we shall give a review of this work; in the mean time we :
commend it to the notice of the profession. The additions by the American editors,
numerous and important."—*Philadelphia American Journal of Medical Science.*

" It strikes us, after a patient examination, that no practitioner who has once had
book in his possession would know how to dispense without it. The editors, or in
authors, appear to have wholly prepared the first part, a most excellent and indispens
addition to the original text. Throughout the entire volume the additions they l
made are readily recognised, and form an essential feature in the construction of
American edition. To students of Medicine especially we recommend this editio
being superior to any other work extant for them."—*Boston Medical and Surgical Jour*

A TREATISE ON THE MEDICAL JURISPRUDENCE OF INSANI
By I. RAY, M. D., 1 Vol. 8vo.

" We have seldom engaged in the performance of a duty, either more agreeable in it
or more gratifying to our pride of country, than that of making our readers acquai
with Dr. Ray's Treatise on the Medical Jurisprudence of insanity; a work, wl
whether we regard it as a contribution to the cause of humanity, or as an attempt to
body the results of moral science, in relation to mental disease and its incidents, is equ
worthy of our admiration." *American Jurist.*

ANATOMICAL, PATHOLOGICAL AND THERAPEUTIC RESEARC
UPON THE DISEASES KNOWN UNDER THE NAME OF GAST
ENTERITE, PUTRID, ADYNAMIC, ATAXIC, OR TYPHOID FEV
ETC., COMPARED WITH THE MOST COMMON ACUTE DISEAS
By P. CH. A. LOUIS. Translated from the original French by HENRY I. B
DITCH, M. D. 2 vols. 8vo.

PATHOLOGICAL RESEARCHES ON PHTHISIS. By P. CH. A. Lc
Translated from the French, with Introduction, Notes, Additions, and an E
on Treatment. By CHARLES COWAN, M. D. Revised and altered by HENR
BOWDITCH, M. D. 8vo.

RESEARCHES ON THE EFFECTS OF BLOODLETTING IN SOME
FLAMMATORY DISEASES, AND ON THE INFLUENCE OF T
TARIZED ANTIMONY AND VESICATION IN PNEUMONITIS. By
CH. A. LOUIS. Translated by C. G. PUTNAM, M. D. With Preface and
pendix by JAMES JACKSON, M. D. 8vo.

ANATOMICAL, PATHOLOGICAL AND THERAPEUTIC RESEARC
ON THE YELLOW FEVER OF GIBRALTAR OF 1828. By P. CH
LOUIS. From Observations taken by himself and M. TROUSSEAU, as mem
of the French Commission at Gibraltar. Translated from the manuscript b
C. SHATTUCK. Jr. M. D. 8vo.

☞ *Law Libraries purchased.*

ABBOTT (Charles, Ld. Tenterden.) Treatise of the Law relative to Merch
Ships and Seamen. 4th Am. from the 5th Lond. ed. by John Henry Abb
with Annotations by Joseph Story, and an Appendix containing the Ameri
Acts respecting the Registry and Navigation of Ships, &c. 8vo.

AMERICAN JURIST and Law Magazine. 8vo. Quarterly, $5 per ann.

ANGELL (Joseph K.) Treatise on the Right of Property in Tide Waters. W
an Appendix, containing the principal adjudged Cases. 8vo. New ed. prep
ing for the press.

ANGELL (Joseph K.) Treatise on the Limitation of Actions at Law and Suits
Equity. With an Appendix containing an abstract of the Statutes of Limi
tions in the several States, Brook's Reading upon the Statute of Henry VI
&c. 8vo.

ANGELL (Joseph K.) Law Intelligencer and Review. 3 vols. 8vo.

ANGELL (Joseph K.) on Adverse Enjoyment. 8vo.

ANGELL (Joseph K.) A Practical Summary of the Law of Assignment. 12i

ANGELL (Joseph K.) and AMES (Samuel.) Treatise on the Law of Private C
porations Aggregate. 8vo. New ed. in preparation for the press.

ANGELL on Water Courses, new ed. 8vo.

BAYLEY (Sir John.) Summary of the Law of Bills of Exchange, Cash Bills,
Promissory Notes; from the 4th Lond. ed., with Notes by Willard Phillips
Samuel E. Sewall. 2d ed. 8vo.

BARBOUR and HARRINGTON's Equity Digest. 3 vols.

CHITTY on Pleadings. 3 vols.

CHITTY on Criminal Law. 3 vols.

CHITTY on Bills. 8vo.

CHITTY on Contracts. 8vo.

CURTIS (George Ticknor.) A Digest of Cases Adjudged in the Courts of Ad
ralty of the United States, and in the High Court of Admiralty in Engla
together with some Topics from the works of Sir Leoline Jenkins, Kt. Judg
the Admiralty, in the reign of Charles II. 8vo.

CURTIS's American Conveyancer. 12mo.

CUSHING (L. S.) Treatise on the Contract of Sale, by R. J. Pothier. Transla
from the French. 8vo.

DAVIS (Daniel.) Civil and Criminal Justice. 8vo.

DAVIS. Precedents of Indictments. 8vo.

DEBATES in the Congress of the United States on the bill for repealing the Law
the Organization of the Courts of the United States, with the yeas and na
&c. 8vo.

GALLISON's Reports. Vol. 2d.

GREENLEAF (Simon.) Reports of Cases in the Supreme Court of Maine, fr
1820 to 1831. 9 vols. 8vo.

HILLIARD (Francis.) An Abridgment of the American Law of Real Prope
2 vols. 8vo.

HOBART (Sir Henry.) Rep. Temp. Eliz. et Jac. I., reviewed and corrected
Edward Chilton. 1st Am. from the 5th Eng. ed. With Notes by J. M. V
liams. 8vo.

HUGHES on Insurance. 1 vol. 8vo.

HOWE (Samuel.) Practice in Civil Actions and Proceedings at Law in Massac
setts. 8vo.

JACKSON (Charles.) Treatise on the Pleadings and Practice in Real Actions,
Precedents of Pleadings. 8vo.

JOURNAL of the Convention for framing a Constitution of Government for Mas
chusetts from the Commencement of their first session, Sept. 1, 1779, to
close of their last session, June 16, 1780, including a list of Members. Publis
by order of the Legislature. 8vo.

KENT (James.) Commentaries on American Law. 4 vols. 8vo. 4th edition.

MASON (William P.) Reports of Cases in the Circuit Court of the United St for the First Circuit from 1816 to 1830. 5 vols. 8vo.

"These Reports comprise the Decisions of Mr. Justice Story on the First Circui the United States, and follow in order after Mr. Gallison's Reports. The decisions r to a great variety of subjects—Constitutional, Admiralty, Personal and Real Law, Chancery, and are characterized by the profound learning, acuteness and thoroughne research, which are such eminent traits of their author. They will bear a favou comparison, in point of Learning and practical utility, with the best volumes of the lish Reports."

MASSACHUSETTS REPORTS. Tyng's Reports of cases in the Supreme Jud Court of Massachusetts, from 1804 to 1822. (Vol. 1st by Ephraim Willia 17 vols. 8vo.

"These Reports embrace the Decisions of the Supreme Court of Massachusetts the time in which they first began to be published until Mr. Pickering commenced labors as Reporter. During a portion of this period Chief Justice Parsons, whose c acter for learning and ability has been so widely diffused, presided on the bench. Reports have sustained a higher reputation throughout the country than these, or l been more extensively cited. The greater part of the present volumes have been sto typed, and have been enriched by the learned Annotations of Benjamin Rand, Esq the Boston Bar."

MAULE (George,) and SELWYN (William.) Reports of Cases in King's Be containing Cases of Hilary, Easter and Trinity Terms 1813; Michaelmas, Hil and Easter Terms 1813-14, and Trin. Mich. and Hil. Terms 1814-15. In 53d, 54th, and 55th years of Geo. 3d. Edited by Theron Metcalf, in 2 vols.

PHILLIPS (Willard.) On the Law of Patents. 8vo.

PHILLIPS on Insurance. 2 vols. 8vo. New ed.

PICKERING (Octavius.) Reports of Cases in the Supreme Judicial Court of Ma chusetts, from 1822 to 1836. 21 vols. 8vo.

RAND (Benjamin.) A Treatise on the Law Relative to Sales of Personal Prop By George Long, Esq. Second American Edition, with Additions. 8vo.

SUMNER (Charles.) Reports of Cases argued and determined in the Circuit C of the United States for the First Circuit. 3 vols. 8vo.

STANSBURY (Arthur J.) Report of the Trial of Judge James H. Peck, on an peachment by the House of Representatives of the United States. 8vo.

"This volume contains a great amount of learning in regard to contempts of c libels, and impeachments, subjects upon which the counsel on both sides exhibit p of very diligent research. It will be a valuable repository of authority and argun in future cases in which the right of courts to punish for contempts may be brough question."

STEVENS (Robert,) and BENECKE (William.) Treatise on Average and Adj ments of Losses in Marine Insurance; with Notes by Willard Phillips. 8v

STORY (Joseph.) Selection of Pleadings in Civil Actions, with Annotations. ed., with Additions by Benjamin L. Oliver. 8vo.

STORY (Joseph.) Commentaries on the Law of Bailments, with Illustrations f the Civil and Foreign Law. 8vo.

STORY (Joseph.) Commentaries on the Constitution of the United States, wi Preliminary Review of the Constitutional History of the Colonies and State: fore the Adoption of the Constitution. 3 vols. 8vo.

STORY (Joseph.) The Same, abridged by the Author. 8vo.

STORY (Joseph.) Commentaries on the Conflict of Laws, Foreign and Dome 2d ed., much enlarged.

STORY (Joseph.) Commentaries on Equity Jurisprudence as administered in land and America. 2 vols. 8vo. 2d ed., enlarged.

STORY (Joseph.) Commentaries on Pleadings in Equity in England and Amer By Joseph Story, L. L. D., Dane Professor of Law in Harvard University. 2d

STORY (Joseph.) Commentaries on the Law of Agency, as a Branch of C mercial and Maritime Jurisprudence, with occasional Illustrations from Civil and Foreign Law. 8vo.

THE EARLY ENGLISH REPORTERS; valuable Works on the Civil Law; Rep of the several States, and the most extensive collection of Law Books to be fc in the United States.

COMMENTARIES ON THE CONFLICT OF LAWS, Foreign and Domestic, in Regard to CONTRACTS, RIGHTS, and REMEDIES, and especially in regard to MARRIAGES, DIVORCES, WILLS, SUCCESSIONS, and JUDGMENTS. By JOSEPH STORY, L. L. D., Dane Professor of Law in Harvard University.

SECOND EDITION. Revised, corrected, and greatly enlarged. pp. 1104.

The following notice of this edition of the above work, from an eminent jurist, has been received by the publishers.

" It is superfluous to call the attention of the profession to ' this remarkable work,' as it is styled by the French reviewers, who justly predicted that it would ' excite the most lively interest in both hemispheres.' The learned author is the first in our day who has in the fullest sense, employed himself, *ex professo*, on International Law ; producing a work, styled, by Mr. Justice Fergusson, of Scotland, ' the most comprehensive and candid in our language, relating to that department of the law.' Its republication in England and on the Continent, in several languages, is the best testimonial of the sense entertained of its merits, by foreign jurists ; as the rapid sale of the first edition is of its estimation by the profession at home. To this second edition, the learned author has added nearly as much matter as was contained in the first; availing himself, with felicitous energy and judgment, of all the sources of International Jurisprudence which the discussions, occasioned by the appearance of his work, have subsequently discovered ; and leaving nothing to be desired on this subject, either by the practitioner or the statesman. To our own country, composed as it is of a family of independent sovereignties, united by the most intimate and frequent relations of trade, commerce, and marriage, yet each separate in its laws and jurisdiction, the work is invaluable. The questions discussed in it, particularly those relating to contracts, remedies, administrations, and the laws of inheritance and succession, are of daily occurrence ; and no American lawyer ought to deem his library complete, where this Treatise is wanting."

REPORTS OF CASES ARGUED AND DETERMINED IN THE CIRCUIT COURT OF THE UNITED STATES, FOR THE FIRST CIRCUIT. By CHARLES SUMNER, ESQ. Vol. 3. Embracing the Decisions of Mr. JUSTICE STORY, from October Term 1837 to October Term 1839.

A YEAR'S LIFE. By JAMES RUSSELL LOWELL. 1 vol. 16mo.

" We take great pleasure in calling the attention of our readers to this volume of poems. It bears the marks of real poetical power and of a highly cultivated power of composition. It is a real relief in the midst of the forced and hackney specimens of what is called poetry with which our periodicals are crowded, to meet for once with a poet whose inspiration is nature and truth. One reading does not satisfy ; we are delighted by another and another perusal, and cannot but take the mood of the author and go hand in hand with him."—*Boston Daily Advertiser.*

LETTERS OF MRS. ADAMS, THE WIFE OF JOHN ADAMS, WITH AN INTRODUCTORY MEMOIR. By her grandson, CHARLES FRANCIS ADAMS. 2 vols. 16mo. 2d edition.

" We cannot undertake to indicate the most attractive portions of what is throughout highly entertaining, or instructive, or both. Now great people and events of the day are brought before the view, and now modes of dress and ceremony are sketched with a distinctness, of which only the female hand is capable."
North American Review for October 1840.

" It will naturally be presumed that this correspondence of an uncommonly sensible woman like Mrs. Adams, who lived in an eventful period of our history, and was personally, and for the most part intimately acquainted with the great men of her times, must be full of interest and instruction ; and so in fact it will be found to be by every reader."
New York Review for January 1841.

OUTLINES OF ANATOMY AND PHYSIOLOGY, TRANSLATED FR
THE FRENCH OF M. MILNE EDWARDS, Doctor of Medicine, Pro
sor of Natural History at the Royal College of Henry IV, and at the Cen
School of Arts and Manufactures in Paris. By J. F. W. LANE, M. D. Bost
1 vol. octavo, with finely executed Wood Cuts.

CONTENTS.

Preliminary Remarks; General Characters of Living Beings; General Cl
acters of Animals; The Functions of Animals and their Organs; Organic
sues; The Functions of Nutrition; The Nutritive Fluids, or the Blood; Ci.
lation of the Blood; Apparatus of Circulation in Man; Mechanism of the Circ
tion; Absorption; Exhalation, and the Secretions; Transudation, or Sanguine
Effusion; Exhalation; Secretions; Respiration; Apparatus of Respiration;
chanism of Respiration; The Influence of Respiration upon the other Functio
Animal Heat; Digestion; Urinary Secretion; Review; Functions of Relati
Nervous System; Sensation; The Sense of Touch; The Sense of Taste;
Sense of Smell; The Sense of Hearing; Light; The Intellectual and Instinc
Faculties; Motions; The Voice; Functions of Reproduction.

SKETCHES OF THE JUDICIAL HISTORY OF MASSACHUSET
FROM 1630 TO THE REVOLUTION IN 1775. By EMORY WASHBU
1 volume. Octavo.

REPORTS OF CASES ARGUED AND DETERMINED IN THE
PREME JUDICIAL COURT OF MASSACHUSETTS. By OCTAVIUS PI
ERING. Vol. 18. (Volumes 19 and 23 in preparation, which will complete
Pickering's Reports.)

A NAVAL TEXT BOOK, containing a series of Letters addressed to the M
shipmen of the United States Navy, on Rigging, Equipping, and Managing
sels. A set of Tables for Stationing in Watches, at Quarters, and for all E
lutions; the Officers and Crews of all classes of Vessels of War. A Na
Gun Exercise, with Plates, for Stationing at the Guns, and Exercising at Q
ters; the Officers and Crews of all Vessels; and a Dictionary of Sea Terms
Phrases. 1 vol. 8vo.

" A work of this description, containing such a variety of matter, cannot bu
useful to all young seamen, in the merchant as well as the public service. Th
is not now any work to which a young seaman can turn to gain information
his profession except Falconer's Marine Dictionary."

A COLLECTION OF THE PLANTS OF BOSTON AND ITS VICINI
WITH THEIR GENERIC AND SPECIFIC CHARACTERS, PRINCIP
SYNONYMS, DESCRIPTIONS, PLACES AND GROWTH, AND TI
OF FLOWERING, AND OCCASIONAL REMARKS. By JACOB BIGEL
M. D., Professor of Materia Medica, in Harvard University, Member of
Linnæan Societies of London and Paris. Third edition enlarged, and cont
ing a Glossary of Botanical Terms. 1 vol. 16mo.

CHARLES ELWOOD: OR THE INFIDEL CONVERTED. By O.
BROWNSON, (Editor of the Boston Quarterly Review.) 1 vol. 12mo.
Extract from the Preface.
" I have embodied in it the results of years of inquiry and reflection; and I h
thought it not ill-adapted to the present state of the public mind in this community.
deals with the weightiest problems of Philosophy and Theology, and perhaps some mi
may find it not altogether worthless."

OR IN ACTIVE PREPARATION,

BY CHARLES C. LITTLE AND JAMES BROWN.
Boston, January, 1841.

I.

COMMENTARIES ON THE LAW OF EVIDENCE. By SIMON GREENLE
L.L. D., Royall Professor of Law in Harvard University. In one volume, £
This work is arranged in Three Parts. The first will treat of the nature, ba
and general principles of Evidence; its general divisions into positive and
sumptive; the theory and doctrines of Presumptive evidence; and those thi
which Courts will themselves take notice of, without proof. The Second I
will treat of the Objects of Evidence; the rules which govern its producti
and the quantity of proof required; including the subjects of primary and seco
ary evidence, and of original evidence and hearsay, together with that of
admissibility of parol evidence to contradict, vary, or explain that which is
writing. The Third Part will treat of the Instruments of Evidence, whet
written or oral; including the subject of witnesses, and the weight and forc
their testimony; with the province and duty of the jury.

II.

COMMENTARIES ON THE LAW OF PARTNERSHIP, by JOSEPH STO
L. L. D., in the course of preparation for the press.

III.

THE RIGHTS, DUTIES, AND OBLIGATIONS OF THE OWNERS, M
TERS, OFFICERS, AND MARINERS OF SHIPS IN THE MERCHA
SERVICE. By GEORGE TICKNOR CURTIS. 1 vol. 8vo.

" Messrs. LITTLE & BROWN,
" Gentlemen—I have read through with great care a large portion of a Treatise
George T. Curtis, Esq. upon the Rights, Duties, and Obligations of the Owners, Mast
Officers, and Mariners of Ships in the Merchant Service. I think the work is writ
with great ability, accuracy, and learning; and if published it will constitute by far
most valuable Treatise now in existence on this highly important branch of law, and ·
be worthy of extensive public patronage.
" I am very truly and respectfully yours, JOSEPH STORY.

IV.

A TREATISE ON THE LAW OF PRIVATE CORPORATIONS. By JOSI
K. ANGELL. 8vo. Second Edition.

V.

A TREATISE ON THE RIGHT OF PROPERTY IN TIDE WATERS.
JOSEPH K. ANGELL. 2d edition. Revised, with a general alteration in the m
ner of treating the subject, and comprehending the English and American de
ions made since the publication of the 1st edition.

VI.

REPORTS OF CASES ARGUED AND DETERMINED IN THE SUPRE
JUDICIAL COURT OF MASSACHUSETTS. By THERON METCALF, [S
cessor to Pickering.] Vol. 1, part 2, which completes the volume.

VII.

REPORTS OF CASES ARGUED AND DETERMINED IN THE SUPREI
JUDICIAL COURT OF MASSACHUSETTS. By OCTAVIUS PICKERING, E
Vol. 9. SECOND EDITION; with Notes and References, by J. C. PERKINS, E

VIII.

A DIGEST of the 17 vols. Massachusetts and 23 vols. Pickering's Reports, c
plete in one volume, octavo, will be published as soon as Mr. Pickering prep
for the press the 19th and 23d volumes, now wanting to complete his Reports.

IX.

THE OLD CHRONICLES OF PLYMOUTH COLONY, IN NEW E
  LAND, now first collected from unpublished manuscripts, and contemp
neous printed documents; illustrated with historical and biographical Nc
By the Rev. ALEXANDER YOUNG, of Boston, Member of the Massachus
Historical Society. With a head of Gov. Edward Winslow, from an orig
portrait painted in 1651.

The value and interest of this work will be greatly enhanced by the fact th
will contain an authentic narrative of the origin and settlement of the Col
written at the time by the first planters themselves. Mr. Young has fortuna
recovered the most important part of Gov. Bradford's lost history of the Plymc
people, and has other documents written by Bradford and Winslow, some of wl
have never been printed, and others are wholly unknown in this country.

This work will be a prior document to Morton's *New England's Memorial*,
will constitute the beginning and foundation of our history. It will make an oct
volume of 450 pages or more.

X.

HISTORY OF THE COLONIZATION OF THE UNITED STATES.
  GEORGE BANCROFT. Abridged by the Author for School Libraries and Scho
2 vols. 16mo.

This Abridgment is designed for the young. .Its object is to put within the re
of the schools of the country a sufficiently full account of the settlement of ev
one of the States. This Abridgment contains not only the history of the Col
zation of the Atlantic States, but of the French settlements from Canada
Louisiana. The attempt has been made to omit every thing beyond the re
comprehension of an intelligent class of scholars, and to furnish an exact, an
possible, an interesting continued narrative, suitable for a class-book in reading
as a manual for instruction.

The work is adorned by illustrations, and will be published in March, 1841.

XI.

PREPARING FOR THE PRESS.

A CONTINUATION OF THE HISTORY OF THE UNITED STAT
By GEORGE BANCROFT. The three volumes already published complete the I
tory of the Colonization. This SECOND PART of Bancroft's History will contain
HISTORY OF THE AMERICAN REVOLUTION, in 2 volumes, octavo, w
rich illustrations. This division of the work will comprise the History of
Country from 1750 to the Peace of 1783.

XII.

BOSTON JOURNAL OF NATURAL HISTORY; containing Papers and Cc
  munications read before the Boston Society of Natural History and publisl
by their direction. Vol. III. No. 4 will be ready in May.

☞ Constantly for sale, a large stock of the most valuable Foreign and Am
can works. Orders for Books will be promptly executed from London .
Paris by the steam packets.

Lightning Source UK Ltd.
Milton Keynes UK
UKHW011958230119
336088UK00017B/1476/P